S0-GQE-961

SOUTH AFRICA

SAIS African Studies Library

General Editor
I. William Zartman

SOUTH AFRICA

The Political Economy of Transformation

edited by
Stephen John Stedman

Lynne Rienner Publishers • Boulder & London

Published in the United States of America in 1994 by
Lynne Rienner Publishers, Inc.
1800 30th Street, Boulder, Colorado 80301

and in the United Kingdom by
Lynne Rienner Publishers, Inc.
3 Henrietta Street, Covent Garden, London WC2E 8LU

© 1994 by Lynne Rienner Publishers, Inc. All rights reserved

Library of Congress Cataloging-in-Publication Data
South Africa : the political economy of transformation / edited by
    Stephen John Stedman.
        p. cm. — (SAIS African studies library)
    Includes bibliographical references and index.
    ISBN 1-55587-421-5 (alk. paper)
    1. South Africa—Economic policy. 2. South Africa—Economic
conditions—1991–   . 3. South Africa—Politics and government—1989–
4. Industrial mobilization—South Africa. I. Stedman, Stephen
John. II. Series: SAIS African studies library (Boulder, Colo.)
HC905.S685   1994
338.968—dc20   93-33330
                                                      CIP

British Cataloguing in Publication Data
A Cataloguing in Publication record for this book
is available from the British Library.

Printed and bound in the United States of America

    ∞   The paper used in this publication meets the requirements
        of the American National Standard for Permanence of
        Paper for Printed Library Materials Z39.48-1984.

All chaps. noted

Contents

Preface

The title of this book reflects my consideration of what needs to be explained if South Africa is to become a long-lived multiparty democracy. As I describe in Chapter 2, transition deals only with the formal institutions of a polity, whereas transformation involves the informal rules, interactions, and power relationships upon which formal institutions rest. While much attention has been devoted to whether South Africa will make the transition to multiparty democracy, less study has been directed at the factors that will determine whether South Africa will sustain democracy. The chapters in the book go far beyond current political developments in South Africa and capture a series of conflicts, problems, and dilemmas that will face that country for the foreseeable future.

Like previous books in this series, this book is the product of an annual SAIS Country Day Conference; this one held on April 13 and 14, 1992. I would like to thank all of the participants of the conference, including the discussants and authors whose comments and papers do not appear in this book: Millard Arnold, Danisa Baloyi, Graeme Bloch, Michael Clough, Chester Crocker, Errol de Montille, Patricia Kiefer, Anthony Marx, Lindiwe Mabuza, Steven Mufson, Sam Nolutshungo, Thomas Ohlson, Witney Schneidman, Gaye Seidman, Timothy Sisk, Rosalind Thomas, and Richard Tomlinson. The conference could not have taken place without the able, cool management of Theresa Taylor Simmons, who also assisted in the compilation of the book. Thanks must also go to four chapter authors who were not present at the conference; to my senior colleague, I. William Zartman, for helping to coordinate the publishing of the manuscript while I was in South Africa; to Gia Hamilton at Lynne Rienner Publishers; and to Victoria White, who checked the manuscript and prepared the index.

The editing of the manuscript took place at the Centre for Southern African Studies at the University of the Western Cape, where I was a visiting fellow from January to September 1993. I would like to thank the Centre and its codirectors, Rob Davies and Peter Vale, for their encouragement and hospitality. Acknowledgments must go to the Fulbright Grant program for a Senior Research Grant that made my stay in South Africa possible. This book is evidence of the depth of the Fulbright South Africa program, as two of the other authors—Amy Biehl and Jeffrey Herbst— were also Fulbright scholars there during 1993, and one other author— Thomas Krattenmaker—was a South Africa Fulbright scholar in 1991.

My final thanks go to my wife, Corinne Thomas, who accompanied me to South Africa and shared the highs and lows of living there. The joys were many—the friendship, creativity, and ingenuity of South Africans of all walks of life, all colors and political persuasions, quietly, often invisibly, working toward a new nation.

The sorrows were few, but profound.

In April of 1993, the very weekend that I finished editing this book and writing the introduction, Chris Hani—a central character in three of the chapters of the book—was assassinated. A crucial link between the leadership of the ANC and its most marginalized constituencies—youth and Umkhonto we Sizwe (MK) cadres—Hani stressed the need for a negotiated transition in South Africa, condemned the violence of the Azanian Peoples Liberation Army (APLA), and called for the establishment of a "Peace Corps" for township youths. As I watched the outpouring of grief and anger at his funeral, I wondered if South Africa's transition could survive his death.

Even with such a setback, the journey toward democracy in South Africa continued. Over the next few months every party, with the exceptions of the Inkatha Freedom Party, the Conservative Party, and a few of the parties of the homeland governments, agreed on a date for South Africa's first inclusive election. The ANC and NP prepared to form a transitional council to steer the country to democracy. I prepared to leave the country in August 1993 cautiously optimistic about the future of South Africa.

Four days before I left South Africa, and only two days before she was to return to the United States to attend graduate school, Amy Biehl—a close friend and author of the chapter on women and democratic transformation in South Africa—died in Guguletu township outside of Cape Town. Amy possessed boundless energy and refused to accept the informal apartheid that continues to separate blacks from whites and rich from poor in South Africa. She was one of the bravest souls I have ever known: she had fears about "crossing the line," as it is sometimes described in South Africa, but overcame them every day.

Two lives then are and will always be associated indelibly in my mind with this book. Both Chris Hani and Amy Biehl, from such drastically different backgrounds, understood and insisted that without a democratic transition, a more fundamental transformation of South African society could not take place. But they also understood and insisted that without that transformation, a democratic transition would not be long-lived in South Africa.

Amy was appalled at the distinct reactions of the media to violence in South Africa. "White deaths," she said, "involve individuals with families, friends, and most importantly, names. Black deaths involve numbers." She once told my wife that if anything ever happened to her in South Africa, she hoped that she would be just another nameless victim. She was too special to too many people for that to happen—and that should be said of all the victims of South Africa's violence.

This book then is dedicated to Chris Hani, Amy Biehl, and all of the victims, named and unnamed, of violence in South Africa, whose deaths pose an incredible burden on the living: that their deaths not be in vain. No better memorial than a living one: a just, peaceful, democratic South Africa, where there are no lines to be crossed or fears to be overcome.

—S. J. S.

1 ─────────────────────

Introduction

Stephen John Stedman

In February 1990, South Africa entered its second interregnum—a period of great uncertainty when old rules had collapsed with no clear sense of what would replace them. Unlike South Africa's first interregnum of the 1980s—a decade of ungovernability when the efforts of the white government to maintain racial supremacy vied with a liberation movement unable to force revolutionary change—many South Africans in the 1990s believed that a transition to majority rule would deliver the country from its political limbo. Accompanying that optimism, however, were heightened expectations and violence as political parties maneuvered to be part of a new political compensation or tried to wreck a nonracial, democratic future.

By December 1993 the shadows, if not the substance, of a new order were at last discernable. A transitional executive council consisting of representatives of different parties was established to oversee South Africa's first nonracial, one-person-one-vote national election, scheduled for April 1994. The poll would choose a constituent assembly and establish a government of national unity consisting of ministers from all parties that received a minimum percentage of the vote. The constituent assembly, bound by prior principles set by the political parties, would write South Africa's new constitution and act as the interim government's legislature. The government of national unity would last for a period of five years as the new constitution was gradually put into effect. At the end of the five years a new election would be held.

This book describes some fundamental economic, social, and political problems that will face the interim regime of national unity that emerges in 1994 in South Africa and the government that follows five years later. The problems are basic, long-term in nature, and defy easy solutions. Nonetheless, if democracy is to be sustained in South Africa, these problems will need to be addressed. Without progress toward transforming the existing economy, society, and state, South Africa's transition to democracy will falter.

Arrangement of the Chapters

In Chapter 2, I define two key words used throughout the book—transition and transformation—and explore the interrelationships between the two in recent works on democratic theory and about South Africa. Based on a distinction raised by Robert Price in his book *The Apartheid State in Crisis,* I refer to transition as changes in a polity's formal institutions and to transformation as changes in the substructure of domination that underpins those formal institutions: cultural norms, social and political values, economic relationships, and resources of key actors. I then ask whether changes that took place during the 1980s in South Africa's substructure of domination provide a firm basis for the transition to stable, multiparty democracy.

Political and social transformations of the 1980s provide an ambiguous legacy for a democratic South Africa. Major changes in South African society rendered that country's racial authoritarianism untenable. But while some changes—especially the rise of a thriving civil society and culture of negotiation—make possible the creation of a stable, multiparty democracy, other changes—intensified economic crisis, intolerance, and violence—could just as easily lead to political disintegration, or what Heribert Adam and Kogila Moodley call "multiracial domination." The result depends on whether elites who enter into narrow transitional pacts are willing to work toward social and economic transformation that can undergird democracy. Such a willingness rests on the ability and efforts of the main political parties, business, and organized labor to create mutual interests through debate and problem solving. Evidence in the last three years of South Africans stepping beyond their limited conception of self-interest has been mixed, but encouraging.

Chapter 3, by Jeffrey Herbst, examines what many believe will be the crucial problem for South Africa's transition: the simultaneous restructuring of South Africa's economy so that it will prosper and the redistribution of wealth and resources to the formerly dispossessed majority. While Herbst acknowledges the imperatives of the latter, he specifically looks at two components of the former: the necessity of fiscal responsibility on the part of the new state and the need for curbing wages. He concedes that the African National Congress (ANC) has shown much sensitivity to questions of macroeconomic balance and investment, but questions whether their good fiscal intentions will be overwhelmed by the need for immediate social spending and constrained by the high-wage policy of organized labor, one of the ANC's most important and best-organized constituencies. He concludes on a pessimistic note, believing that the ANC will not have the political will to carry out policies that will generate much needed economic growth.

In Chapter 4, Colin Bundy investigates the crucial role that will be played by youth in South Africa's future. South Africa's democracy must find a positive role for youth if it is to survive in the future. In one sense, this is simply an acknowledgment of demographic realities. But deeper than the question of numbers is the question of the political attitudes of youth. In the 1980s, "the children of iron" were key actors in making South Africa ungovernable. Deprived of childhood, thrust into a day-to-day political struggle, and given the enormous power of "deciding who would live and who would die," radical youth inspired both admiration and fear among their elders.[1]

No section of South African society has faced a greater challenge in the transition from revolutionary struggle to the politics of democratic negotiation than youth. Whereas the ANC's ex-guerrilla fighters have the benefit of militaristic discipline and leaders who were given prominent positions in the ANC's executive committee, South Africa's radicalized youth have been largely ignored by ANC leaders. The lack of inclusion in the transition negotiations has led the ANC Youth League to oppose the power-sharing interim arrangements between the ANC and the National Party (NP) and to experience a larger alienation from the creation of a new South Africa. Peter Mokaba, head of the ANC Youth League, has warned that youth feel disenfranchised from the struggle to which they had contributed:

> No one can do anything properly without the youth. We say that for negotiations to succeed they must involve all the people who will be affected. . . . We agree that there should be negotiations but the question bugging the minds of youth and others is whether negotiations are about the things we have been fighting for.[2]

Mokaba warns that if the needs of youth are not addressed in a future democracy, they will pose "a greater threat to SA's stability than either the Right or a reactionary bureaucracy."[3]

Democracy in South Africa will also depend in large part on whether its leaders nurture and sustain connections to citizens at the grassroots level. In South Africa such connections may prove easier to maintain because of the growth of important components of civil society such as civic associations, labor union membership, and peace committees over the last eight years. In Chapter 5, I. William Zartman explores the connections between local-level negotiations, transformation of society, and democracy in South Africa.

As Zartman observes, local-level negotiations can prove crucial for democratic transformation in South Africa. First, the participation in local-level politics can empower citizens, who can then hold rulers accountable for their actions. Second, as Frederik Van Zyl Slabbert eloquently describes, "Important as national negotiations about democratic transition are, negotiations at the regional and local level can be critical in restoring and healing the fractured civil society of South Africa."[4] Zartman warns, however, that the effectiveness of local negotiations in assisting democratic transformation depends on whether they are incorporated into a national process or are ignored, controlled, and repressed in a national pact.

In Chapter 6, Amy Biehl looks at the need to transform the position of women in South Africa. Although women make up more than half of South African society, they are seriously underrepresented in positions of economic and political power. If democracy is to have a tangible impact on South Africa's people, it must address the problem of substantive equality for women. But the imperative to empower women in South Africa is not only normative; it is also practical. Like local civic organizations, women can be a powerful lobby for accountability in South Africa and play a leading role in forcing government to carry out democratic praxis and providing democracy with the legitimacy it will need to survive.

While Biehl puts forward various constitutional and legal remedies that can assist in transforming the position of women in society, Thomas Krattenmaker in Chapter 7 poses a series of dilemmas that will face any South African government that tries to use legal remedies to bring about social change. While Krattenmaker looks only at racial law, his examples from the United States show the delicate balance that must be achieved in crafting laws that address discrimination, yet maintain the basic privacy of individuals. Although Krattenmaker does not say so directly, he suggests that the use of legal remedies to address discrimination can have the net effect of racializing a struggle that the ANC has maintained is a nonracial conflict.

In Chapters 8 and 9, Herbert Howe and Jacklyn Cock respectively address the role of security forces in a democratic South Africa. The issue is crucial, for the military under successive apartheid regimes was used to carry out a covert war against critics of the regime within South Africa and in the larger region. Both of the authors focus on the need to bring together the various warring factions in the country—the South African Defence Force (SADF), Umkhonto we Sizwe (MK) guerrillas, and homeland armies. While Howe argues that integration will take place on SADF terms, Cock cautions that integration without transformation of the norms and values of the South African Defence Force will prove counterproductive to strengthening the larger democratic transition.

Chapters 10 and 11, by Peter Vale and Gilbert Khadiagala respectively, provide the regional context for South Africa's transition to democracy.

Vale presents a wide array of crises and threats to regional security that can undermine progress toward democratic stability in South Africa and in its neighbors who are also building democratic institutions. He believes that South Africa can play a key role in assisting its neighbors to find regional solutions to these crises and threats, but such assistance will be forthcoming only if the new South Africa transforms its diplomacy and foreign service. The crux of the matter is whether South Africa can create a new self-image that it can project in the region and in the larger world.

Khadiagala asks a different question: "What will the simultaneous transitions underway in the region mean for regional identity and cooperation?" He contends that the turn to multiparty democracy in the region will bring new actors with new priorities to the national forefront. Khadiagala argues that the domestic transitions could bring to an end much of the regional cooperation that was present in the 1980s. A democratic South Africa will provide the region's other states with a dilemma: Should they act collectively to bargain as one to compensate for South Africa's asymmetrical power in the region, or should they negotiate bilaterally with South Africa? The example of Frederick Chiluba in Zambia, Khadiagala suggests, provides evidence that patterns of interstate relationships in the future in Southern Africa will be functional and bilateral. This, says Khadiagala, is not necessarily bad: the region would be better served by pragmatic policies that bolster the current domestic transitions, rather than by regional integration schemes that are likely to come to naught.

Chapter 12, by Robert Price, concludes the book. He rejects the view that conflict over political identity will be paramount in the future in South Africa and insists that economic conflicts over growth and distribution will be central. Price is optimistic about South Africa's democratic future. He argues that the transformations of the 1980s provide the basis for a thriving civil society, a key to democratic consolidation. But Price warns that if South Africa's leaders choose constitutional principles that address identity rather than distribution, South Africa's prospects for democracy will be diminished. For economic growth to bolster South Africa's democratic prospects, a strong central government will be needed to enter into partnership with organized labor and business to restructure South Africa's economy. Price ends his essay by speculating that the failure to address the key questions of growth and distribution could lead South Africa to the experience of Weimar Germany in the 1920s.

Upon reading this book, one may come away pessimistic about South Africa's ability to meet the overwhelming challenges of building a stable democracy. But one should not underestimate the abilities and talents of South Africans to solve their problems. These essays should be read in the spirit of attempting to contribute to a dialogue within South Africa, a competition among ideas that has already led to creative, ingenious attempts to resolve their myriad conflicts—a process all of us hope and expect will continue to bear fruit.

Notes

1. "The children of iron" comes from J. M. Coetzee's apocalyptic novel of South Africa's first interregnum, *The Age of Iron* (New York: Random House, 1990), pp. 48–51. The power of "deciding who would live and who would die," comes from Mamphele Ramphele, "Social Disintegration in the Black Communities—Implications for Transformation," *Monitor.*

2. Quoted in the *Financial Mail*, March 5, 1993, p. 48.

3. Ibid.

4. Frederik Van Zyl Slabbert, *The Quest for Democracy* (London: Penguin, 1992), p. 93. Slabbert dedicates this book to those he has worked with on the Metropolitan Chamber in greater Johannesburg for helping him "to understand how essential services like water, electricity, sewerage and refuse removal can help to sustain democratic practice" (p. iv).

2

South Africa:
Transition and Transformation

Stephen John Stedman

The foundation upon which the formal political order rests can be thought of as a "substructure" of domination—social interactions, cultural norms, economic activities, and informal power relationships that create the basis for compliance with the prescriptions of the ruling group. Changes in the underlying structure are often the precursor of and condition for alteration in the political system's "superstructure," the formal system of power. Basic alterations in the former, the substructure of power, can be thought of as involving political transformation, and can be distinguished from political transition, the movement from one formal arrangement of power to another. Transformation prepares the way for transition.

—Robert Price, *The Apartheid State in Crisis*

Trying to analyze the potential for democratic transition in South Africa has been likened to trying to check the oil level in a car whose engine is running. The shaking of the motor and chugging of the pistons obliterate any fixed mark from which to judge the motor's worthiness. Nonetheless, pundits and scholars alike keep putting their dipsticks into the South African sludge and keep believing that their readings are accurate. Some see the oil level as adequate and pronounce South Africa's democracy as ready to roll; others fret about putting the engine into gear under so much uncertainty. Of course, the proof in this case will come only when someone tries to drive the vehicle.

If the study of transition can be likened to examining the motor of political change, the study of transformation can be compared to analyzing the chassis in which the motor sits. Here political analysts can make firmer judgments: even if the motor is purring, the car will not go anywhere if it has three flat tires and lacks a steering wheel. But in the case of South Africa, social scientists tend to hover under the car's hood, poking at the plugs, points, belts, and hoses of transition, and never check to see if the car is sitting on cinder blocks or steel-belted radials.

Such transition fixation reflects the rise of a narrow school of analysis that believes that democracy depends solely on agreements made by elites, and that democratic transition will naturally result in democratic consolidation and stability. This chapter opposes such a view and argues that while the agreement of elites is necessary for democracy to emerge, it is insufficient to sustain and consolidate democracy. For that to take place, elites who enter into transition must actively labor to transform their societies and economies to provide the pillars, or to keep in metaphor, the working chassis for democracy. In South Africa, such a transformation will not be easy, but it is by no means impossible.

Narrow Transitionism

For some current analysts of democracy, when South Africa has its first nonracial election, the story will end there. Having agreed reluctantly to become democrats, elites of different political persuasion need only to remain committed to their agreement for democracy to succeed. The most extreme example of this school is Giuseppe DiPalma's work on transitions to democracy, which dismisses the consolidation of democracy as a trivial process and states baldly, "The decisive role in establishing democracy belongs to the agreement phase, not to consolidation."[1] In a similar vein, by focusing on democracy solely as transition, the work of Terry Lynn Karl asserts that there is no precondition necessary for democracy to emerge and that "very uncivic behavior, such as warfare and internal social conflict" rather than "trust and tolerance" can lead to democracy.[2]

Two arguments, one empirical and one normative, provide the bases for the narrow transitionists to claim that there are no essential preconditions for democratic success. First, many countries in Latin America and Africa have embarked on formal transitions to democratic institutions. Many of these countries suffer from desperate poverty and seemingly disprove the claims that economic development is necessary for democracy to emerge. Second, from a normative perspective, a belief in absolute preconditions would lead one to ignore these new experiments in democracy. Lacking essential economic and social prerequisites, they are bound to fail. The absence of those prerequisites can then justify the most horrific authoritarian practices of governments that preside over societies not developed enough for democracy. On the other hand, the belief in no preconditions justifies the demands for harsh economic conditionalities and rigid structural adjustment programs and allows the narrow transitionists to ask the impossible of the new experiments in multiparty democracy.

These considerations lead the narrow transitionists to use questionable data and logic. Take, for instance, René Lemarchand, who lambasts the work of Almond and Verba on civic culture, Lipset on social requisites of democracy, and Rustow, who put forward national unity as a necessary background condition for democracy:

> If national unity is to be regarded as a basic precondition, what is one to make of Mauritius, which is surely one of the most culturally fragmented societies anywhere in the world—and one where democracy has shown remarkable vitality. Or take the cases of Benin, Senegal, or even Zambia, where the levels of poverty are among the most grinding on the continent: fragile as their respective holds on democracy may be, these examples are likely to erode further still what little credibility the Lipset thesis still has among scholars. As for the "civic culture" or "five Cs" argument, one is compelled to wonder why it might conceivably apply to countries like Senegal, Botswana, or Mauritius, and not to others.[3]

This paragraph displays a core assumption of the narrow transitionists: Short-term success at establishing democratic institutions is tantamount to having established consolidated democracy. Leaving aside the remarkable cases of Mauritius and Botswana, which have sustained democratic institutions for over twenty-five years, Lemarchand's assault on democratic preconditions is based on the experiences of Senegal, a de facto one-party democracy for fifteen years, and Benin and Zambia, both of whose democratic experiments are less than two years old and have already shown signs of backsliding. The fragility of democracy in these latter cases matters: next year, none may have formal democracy.

If the point is simply to say that there are no preconditions for democracy to emerge, then these cases can be used to disconfirm a categorical cause and effect proposition. But should it be surprising that in the complex, chaotic, indeterminate world we inhabit, no single cause can bear the burden of explaining the complex historical process of the development of democracy? Most social scientists who take their self-identification seriously would put the question in probabilistic terms: All things being equal, is democracy more likely to succeed when there is economic growth, a vibrant middle class, national unity, and a civic culture?

The alternative theory of the narrow transitionists is that democratic institutions themselves create economic growth, equitable income distributions, and a civic culture. In the words of Karl, these good things "*may* better be treated as the products of *stable* democratic processes, rather than as the prerequisites of their existence."[4] The crucial words here are "may" and "stable," which prompt two questions. In situations where a narrow transition has taken place in unamenable circumstances, *can democracy survive if it does not lead to economic growth, lessen inequalities, and create national unity and norms of tolerance?* Second, *in the absence of these good things, how likely are democratic processes to be stable?*

Democracy may lead to all good things, but it certainly does not do so automatically. By sweeping aside insights and contributions from sociology, economics, and history about factors that make probable the consolidation of democracy, the narrow transitionists breed complacency about the hard work that is needed to sustain democracy once an elite agreement is in place.

The literature on the social and economic requisites of democracy (which should be recast in a probabilistic and not deterministic form) insists that at some point there must be a connection between politics at the top and the broad social, economic, and demographic forces that affect people's lives in society writ large. One can agree with the narrow transitionists that political democracy is not economic democracy, that it is not social democracy. One can also agree that political democracy, narrowly defined, is a good in and of itself. The important issue in South Africa and the nations of the world's periphery, however, is whether political democracy can survive if it doesn't lead to progress toward economic and social democracy. And on this question the narrow transitionists have been rather silent.[5]

The case of South Africa can help further our understanding of the difference between democratic transition and consolidation. The narrow transitionists' accounts of democracy have ignored the need for societal and economic transformation to underpin transitions if they are to be sustained in the long term. In the rest of this chapter I ask four questions of the South African case that have relevance for the larger literature on transition: (1) How do changes in South African society generated during the conflict to end white rule in the 1980s relate to conditions necessary to sustain democracy? (2) What barriers stand in the way of consolidating democracy in South Africa? (3) Will the terms of the formal transition to democracy in South Africa prove to be a stepping stone or stumbling block to the transformations needed to sustain democracy? (4) And finally, what processes can drive elites to move beyond formal rules and build the pillars that will sustain South African democracy?

South Africa in the 1980s and 1990s: Slouching Towards Democracy?

Key transformations of South African society and economy in the 1970s and 1980s created a disjunction between South Africa's formal political institutions of apartheid and herrenvolk democracy and the substructure of domination on which those institutions once rested.[6] Will democracy as a set of formal institutions bring South Africa's political system into proper alignment? The answer to this question depends largely on the analysis of South Africa's societal, economic, and political transformations of the 1980s.

The insurrections of the 1980s in South Africa resulted from a long chain of causation running back to the late 1940s and the implementation of apartheid. National Party policies of state patronage moved most Afrikaners out of rural poverty and into bureaucratic jobs. Economic growth, in part the product of the government's protectionist economic policies, dramatically raised the living standards of Afrikaners. By the 1970s, however, the successful performance of South Africa's economy demanded the relaxation of formal prohibitions on labor mobility and collective bargaining. The very success of apartheid in lifting Afrikaners into affluence induced divisions within their ranks concerning the need for political change and created an opening for the National Party in the early 1980s to attempt basic reforms. Those reforms, however, had the unanticipated effect of provoking black resistance in the townships. Instead of creating a stratum of blacks who would accept white supremacy in exchange for material prosperity, townships were rendered ungovernable by a coalition of youths, labor, church members, and women who looked to the African National Congress in exile as their legitimate leaders. In turn, the ungovernable townships were then themselves the basis for the establishment of dual authority as street committees, civic associations (civics), and people's courts replaced the discredited structures of the apartheid state.[7]

By the late 1980s, the formal political rules in South Africa had no legs to stand on. New political institutions were therefore needed to harmonize the economy, society, and polity of South Africa. But are the changes, internal conflicts, and uncivic behavior of the 1980s likely to lead to democracy in the 1990s and beyond, as Karl suggests above, or are they just as likely, as others speculate, to lead to one-party hegemony or reversed racial domination?[8]

In his conclusion to this book, Price answers that the changes of the 1980s, especially the rise of civic associations and local authority structures, will play a key role in creating a vibrant civil society—a crucial pillar necessary to consolidate multiparty democracy. Others, however, question whether arrangements forged during a struggle for liberation and revolution can adapt and contribute to a different historical context.

South Africans, not just from the liberal community, argue that for the civics to play a positive role in civil society, they must stand outside formal political structures. The ANC and the South African National Civic Organisation (SANCO) acknowledge the need for a strict separation between political parties, a new formal authority, and extraparliamentary opposition. But this demands the "T" word: a transformation in identities and strategies that must cope with new and unanticipated challenges. So far, according to Jeremy Seekings, the results have been meager. Many civics leaders harbor explicit national ambitions, and tensions have arisen at grassroots levels between the ANC and civics on the division of responsibilities between politics and development work. Finally, a shortage of resources has hindered both the civics and the ANC from making necessary changes to support their new official roles.[9] And some analysts have raised the question of whether financial support for the civics will vanish once an internationally legitimate government is in place. Flows of cash from nongovernmental organizations (NGOs) and international lending organizations can skew the balance of power away from the grassroots to the central government.[10]

The crunch time for the civics will come, Seekings suggests, when the first elections take place. Elected officials

> will claim a superior mandate, and will take over much of the developmental role played by some civics now in the absence of legitimate and representative local government. Any claim by a civic to be the "true" representative of the "community" will be contested by elected councilors.[11]

Tensions between civics and the new government could just as easily lead to ungovernability as to democratic accountability.

Khehla Shubane raises another issue concerning the ability of civic associations to play a positive role in civil society: whether they will possess political tolerance.

> To promote democracy civics will have to start by learning how they themselves can become more tolerant. Quite simply this is a question of how prepared civics are to live with other organizations in the community, organizations which may be organizing the same community.[12]

The issue of political tolerance within the civics came to the fore in South Africa in February 1993 when Dan Mofokeng, SANCO's general secretary, announced that white parties would not be allowed into townships to campaign in a new election because of their previous affiliation with apartheid structures. While the ANC criticized this stance, the announcement revealed a tendency among some civics to try to be a gatekeeper and control information to their communities.[13]

Another major difficulty is that the present civic associations seldom represent all interests in a community. When a civic attempts to be a gatekeeper either to outside development agencies or to competing political interests, conflict tends to erupt between the civic and those members of the community who feel that the civic does not represent them.[14]

This leads to the broader issue of attitudes of struggle and liberation, which may not be compatible with the informal norms necessary to foster multiparty democracy. Donald Horowitz, for instance, in his book *A Democratic South Africa?* sees the politics of liberation as a hindrance to South Africa's democracy, because it has raised expectations, destroyed tolerance of opposition, created an antidemocratic style of organization, and eschewed the compromises necessary for multiparty democracy to thrive. In his words, "Revolutions require the use of undemocratic methods and the consolidation of undemocratic leadership. Both are inimical to the development of democracy later."[15]

The overblown expectations of the resistance in the 1980s may prove problematic to sustaining a democratic transition. Similar to Jeffrey Herbst's analysis in the next chapter, Heribert Adam and Kogila Moodley stress:

> One legacy of the extensive politicization in South Africa is that people now have high expectations that democracy will bring quick material improvements. A new ANC-led government may find itself the victim of the same unrealistic mass expectations that the ANC and other groups encouraged during the period of repression.[16]

The fears of expectations surpassing democratic performance has even led Nelson Mandela to state that the ANC must educate its supporters to expect less of political change.[17]

But the legacies of resistance demand a more nuanced reading than Horowitz provides. Whereas Horowitz asserts that the ANC is a coherent movement suffused with revolutionary intentions, he ignores an important change during the 1980s: the development of a culture of bargaining within the top of the union movements and within the ANC itself. Such a culture of bargaining—the willingness to engage in give and take—can take place only if participants enter negotiations from a position of perceived equality. Here Adam and Moodley rightly point out that the liberation protest of the 1980s has created attitudes necessary for democratic transition: "The dominant mindset of active, resilient protest rather than passive acceptance of subordinate conditions" is "an important precondition of successful negotiations and pacting, and perhaps even a minimal sense of common nationhood."[18]

Horowitz's depiction of the ANC as a revolutionary organization that desires hegemony in the new South Africa presents a caricature. The simple truth is that the ANC does not stand for one thing, is a large tent for diverse constituencies, and is as much distinguished by its pluralism as by its dogmatism. All of these qualities are associated with democratic political parties and provide hope for democratic consolidation in South Africa.

At the elite level, the most in-depth study of tolerance in South Africa concludes that leaders of the ANC–South African Communist Party (SACP) alliance show high levels of political tolerance, scoring only second to members of the Democratic Party (DP). While the leaders of those two parties show tolerance ratings of 75 percent and 83 percent respectively, the National Party (NP), Conservative Party (CP), Indian parties, and Inkatha Freedom Party (IFP) all score less than 50 percent.[19] Contrary to the arguments made by Horowitz and others who see the parties of liberation of the 1980s as the font of intolerance, the survey shows that the parties that were either part of the government, to the right of the government, or compromised by the government harbor the most intolerant attitudes that can threaten stable democracy in South Africa.

The continued high levels of intolerance on the part of actors associated with the government or to the right of the government pose major problems for democratic stability in South Africa. Here one has to return to Price's analysis and ask whether he fully captured the transformations of the 1980s. Both Patrick McGowan and Kate Manzo, and Hermann Giliomee have questioned the extent to which Afrikaner commitment to their identities has changed.[20] Large sections of the white population still score high on racist attitudes.[21]

Another aspect of the transformation of the 1980s concerns lasting effects of the unintended consequences of both government and revolutionary policies. While the government has officially changed its policy of "total onslaught" of the 1980s, many of the individuals associated with the policy are still in positions of influence. Having created a covert government capability to kill and destabilize, the current leaders of the NP have found it difficult to close it down. On the liberation side, opinion polls show an overwhelming acceptance of negotiation and rejection of violence, yet there still remains a hard-core cadre among the youth whose lives have been steeped in violence and who have been unable to change identities.

What then can be said of the transformations of the 1980s? In sum, the transformative changes of the last decade leave a mixed legacy. There should be little doubt that changes in society and economy have created an opening for a democratic transition. But if elite agreement leads to formal democratic institutions, there will still be a profound disjunction between South Africa's formal political institutions and their societal foundations. The mixed legacy of the 1980s will mean that South Africa's democratic consolidation will depend on the ability of its elites to address a wide range of problems and build firmer supports for democracy in "the substructure of domination."

Other Likely Barriers to Democratic Consolidation

Analysts suggest that in addition to the legacies of the 1980s, three other barriers may be paramount in subverting a democratic future for South

Africa: economic legacies of apartheid, ethnicity, and the desire for hegemony on the part of the ANC.

Economic Legacies of Apartheid

A wide consensus among scholars holds that South Africa's skewed economic system and its inequalities will pose the greatest threat to democracy. Echoing Price's analysis in this book, Adam and Moodley insist that the "chances of a future South African democracy and stability do not falter on incompatible identities but depend mainly on the promise of greater material equality in a common economy."[22] Similarly, Frederik Van Zyl Slabbert writes, "Development without democracy is possible, but democracy without development has a very short lifespan."[23] And Timothy Sisk, one of the most impressive U.S. scholars writing on South Africa, states, "Without economic and social transformation, any political settlement will be ultimately unsuccessful."[24]

The deep inequalities between black and white in South Africa will be difficult to resolve because the economy faces a crisis of productivity. If it was the case that the South African economy was healthy and growing at a steady rate, the demands for redistribution could be more easily met. As Stephen Lewis aptly describes, "It is easier to deal with difficult distributive issues . . . if the total size of the cake is increasing."[25] But increasing numbers of analysts now recognize that for South Africa's economy to grow, it will need dramatic restructuring to become internationally competitive.

I will not dwell on the simultaneous challenges of redistribution of resources and economic restructuring to promote sustainable economic growth, as they are addressed by both Herbst and Price in this book. But it should be noted that pessimism about the ability of democracies to make difficult choices in restructuring their economies has prompted some analysts to believe that democracy is doomed to fail in South Africa. A prominent example is the Nedcor/Old Mutual forecasting exercise that, by including in its definition of successful transition to democracy the criterion of rising incomes, comes to the tautological conclusion that "there have been no successful transitions without strong economic performance before, during, and after the transition."[26] Such an analysis can provide a justification for authoritarian solutions to South Africa's crisis.

Ethnicity

Perhaps the most controversial question in South Africa is the extent to which ethnicity may hinder the development of stable democracy there. With the exception of two outside observers of the country, Donald

Horowitz and Arend Lijphart, most analysts believe that ethnicity has been blown way out of proportion as a present or likely future factor in South African political life.

As Price suggests in the conclusion of this book, most black South Africans identify themselves as South Africans before any racial or ethnic category. Only one political party in South Africa, the Inkatha Freedom Party, mobilizes on the basis of ethnicity. Even within Natal and the KwaZulu homeland, where Inkatha is strongest, the IFP currently polls only about 40 percent support. Outside of Natal in the Vaal triangle, conflict has reflected a materially and politically based struggle between "the have-nots" of the hostels, who happen to be Zulu, and those "haves" who live in the communities around the hostels.[27] Violence rose precipitously after the IFP began to mobilize there in 1990.

The danger in South Africa is that what began as politically inspired conflict over material issues can easily lead to indiscriminate hatred and stereotyping of people based on ethnicity. For instance, in June and July 1993 two articles in the *Weekly Mail* described how within the townships around Johannesburg verbal and physical assaults were common on Zulu speakers, as residents assumed such individuals were Inkatha supporters or troublemakers.[28] Once such enemy formation begins, and groups believe that the basis of conflict is endemic to the character of one's adversary, peacemaking becomes extraordinarily difficult.[29]

Horowitz writes that the ANC, while asserting South African identity, is essentially a Xhosa-based party. Yet as Adam and Moodley observe, "there is no evidence that Xhosa culture has been elevated to an ethnocentric ideal" within the ANC.[30] If Horowitz is correct, one can only assume that the anomalous support of massive numbers of non-Xhosa (for instance, 80,000 registered members in Natal alone) is a product of large-scale false consciousness, which Horowitz as an outsider can observe.

The ANC has bent over backwards during the constitutional negotiations in South Africa to acknowledge the symbolic importance of ethnic cultural identity. Some, however, feel that the ANC should do more and include within the constitution "a guaranteed right to secede."

The danger, of course, is that a right to secession can become a weapon to any leader whose support is based on ethnic mobilization. In the case of Natal, where the people are clearly split in their allegiances toward Buthelezi and Inkatha and the ANC, the right to secession can substitute a regional civil war for a national civil war. Moreover, to put the onus of reconciliation firmly on the shoulders of the ANC misses a key aspect of the conflict equation. Donald Rothchild has pointed out that ethnic conflict in African societies can result as much from the nature of demands made by groups as from the response of the government or rivals.[31]

Future ANC Hegemony

Horowitz also contends that a key barrier to a future democratic South Africa is the desire of the ANC to rule as a one-party state. His belief stems from two sources: the triumph of one-party states elsewhere in Africa and the desire for radical transformation among important constituencies within the ANC. I have already questioned the stereotypical view of the ANC as a party bent on domination, so I will focus on the relevance of Africa's experience with democracy as a possible harbinger of South Africa's fate.

Adam and Moodley argue that four reasons mitigate the potential for South Africa descending into "a pseudodemocratic patronage system . . . characterized by high levels of corruption and little democratic accountability."[32]

First, because any new government will have to depend on existing civil servants and bureaucrats, Africans will gradually enter the state and gain experience in good government as a result. Second, "the employment capacity and opportunities in private business for skilled blacks will relieve the civil service from a wasteful expansion." Third, a strong private sector economy will not allow any "post-apartheid government . . . to sabotage the prospects of economic growth through poor governance." And finally, the sheer diversity of social forces in South Africa mitigates the possibility of authoritarianism.[33]

If Horowitz overstates the case for likely one-party hegemony, Adam and Moodley underestimate its potential in South Africa. They naively believe that democracy per se will end patronage politics in South Africa. Their description of forces in society that will insure good governance seems sorely at odds with South African reality. The existing civil service in South Africa has been marred by a series of recent scandals and corruption.[34] South African big business has likewise suffered a systematic decline in ethics.[35] Indeed, the present examples of government and business behavior in the face of crisis present a different example to black South Africans: selfishness and immorality are proper. Finally, much of Adam and Moodley's analysis assumes a healthy, growing economy, which is far from certain.

Transition: Stepping Stone or Stumbling Block?

A theme that runs through this book and that has been emphasized by others is that decisions taken during the transition in South Africa will either make it easier to carry out transformation or will close off possibilities for transformation.

18   SOUTH AFRICA

The question of the transition's role in transformation has two aspects. The first concerns the institutions that are chosen: Will they enable South Africa's rulers to resolve their conflicts or will they prevent them from doing so? The second concerns the process of transition: Will the ways in which the leaders interact provide a positive experience for the future or will they undermine a democratic future?

On the first question, Price, in this book, argues that any constitution that emerges during the transition must provide a central government with enough power to manage the economy, carry out its restructuring, and implement policies of meaningful redistribution. Adam and Moodley agree, but they also recognize that ethnic identity can play a role in disrupting South Africa's transition to democracy and therefore favor a federal solution to South Africa's constitution. They insist that federalism in South Africa must enable a strong center to redistribute resources among unequal regions and to formulate a coherent national economic strategy.[36]

In order for South Africa to overcome the antidemocratic forces of ethnicity and hegemony, Horowitz suggests that the following institutions should be written into a new constitution: a presidential system, federalism, and an electoral scheme called the alternative vote. Important, but less important than the substance of a new constitution, is the process of creating a new constitution; a new constitution should not be the result of narrow "horse trading" among the parties, but the product of inclusive, enlightened deliberation.[37]

Horowitz has drawn much criticism because only about nine of his book's 282 pages address economic problems of distribution. The lack of attention given to that subject has prompted many observers to write that Horowitz has ignored the problem altogether. Yet Horowitz admits that the striking disparities in income, wealth, and resources will pose problems for democracy in South Africa and sensibly cites Robert Dahl's work that meeting some of the demands for redistribution can appease the have-nots. To Horowitz's credit, he recommends a presidential system of government for South Africa, precisely because a strong central government will be needed to deal with the distribution crisis.[38]

Two key questions arise, however. First, will Horowitz's other recommendations dilute the strong central government needed to address the distribution problem? Second, does Horowitz's emphasis on ethnicity and likely ANC hegemony so dominate his analysis that people fasten on to his calls for strong checks and balances to combat those problems and ignore his muted calls for a strong center?

Price, in his concluding essay in this book, answers affirmatively to these questions, whereas I believe in a mixed verdict. To the first question, there is nothing inherent in federalism, for example, that would diminish the center's power to cope with economic crisis. Germany's federalist system has been able to combine strong local representation with a national corporatist framework for business-union-government cooperation in economic policymaking. Indeed, for federalism to work in South Africa, where the constituent units will vary widely in their socioeconomic viability, a strong center will have to play a redistributive role. Yet Price is correct in worrying that some forms of federalism, and certainly the consociational schemes put forward by some, could impinge upon the ability of a new government to carry out the transformation of South Africa's economy.

On the second question I agree with Price, that one of the consequences of Horowitz's preoccupation with and prioritization of the identity and hegemony problem in South Africa has contributed to the delegitimization of the need for a strong state and central government in South Africa. The diagnosis of South Africa's ills plays into the hands of those South Africans who desire what I call a Gulliver state—a massive giant immobilized by numerous constitutional strings.

Will South Africa's Leaders Do the Right Thing?

Let's assume that we can describe a set of optimal institutions and policies that will increase the chance of sustaining democracy in South Africa. What is the likelihood of elites choosing those policies?

In South Africa new institutions and new policies will result from the individual interests of the political parties and their mutual desire to create a new South Africa. As Horowitz argues, however, the institutions and policies that may be the best for the political parties there may not be the best for the public. Indeed, in the case of South Africa the very commitment of the different political parties to democracy is in question. What then will determine which institutions and policies are established?

Horowitz argues that democratic commitment could come from five possible routes: (1) habituation, where actors come to accept democratic norms through their public mouthing of them; (2) deterrence by forces committed to democracy and the establishment of institutions that will strengthen those forces; (3) reciprocity between forces committed to democracy and those who are not—in essence buying off opposition; (4) learning, whereby through the gaining of new information forces not previously committed to democratic institutions become "enlightened"; or (5) a coalition structure that pushes previously uncommitted parties into the moderate middle against the extremes.[39]

Horowitz rightly dismisses the first route—habituation can only come well after democratic institutions have been created. He further argues that deterrence and reciprocity are unlikely because any party with hegemonic aspirations will fight constraining institutions before they are established. This implies that parties can see the distributional consequences of institutions and will fight the creation of arrangements that disadvantage them. Horowitz also dismisses the possibility of actors undergoing a cognitive change by "learning just how explosive and inimical to everyone's interests intergroup conflict and violence can be in a divided society." He contends that South Africa's political history of white exclusive democracy and the high stakes in a new constitution will prevent such a cognitive shift.[40]

Horowitz argues that one should focus on the interests of the political parties and the incentives that may drive them to choose institutions that are in the public good. Here he contends that incentives can change during the transition process as parties differentially commit themselves to democracy. Those actors who first publicly commit themselves to democratic competition will find that they have a common interest against extremists on both sides who are opposed to democracy. The structure of the coalition—a committed democratic middle against antidemocratic extremes—constrains the middle parties to act democratically. In this model, parties with hegemonic pretensions who find themselves in the middle discover that as they defend the democratic process, their interests change incrementally.[41]

Horowitz hopes that the result of such a change will lead the parties of the middle to choose institutions that will best serve the public. He suggests that this will be more likely if the development of a constitution occurs through a "social contract mode" rather than through negotiation. Such a mode emphasizes broad participation and consensus, rather than "horse trading" between narrowly interested parties.[42]

Horowitz's formulation ignores that a coalition of the middle will still face incentives against enlightened policymaking. The first problem, which he mentions himself, is that the parties of the middle can choose to be narrowly exclusive and establish a condominium on power—a "pact"— that omits other powerful actors.[43] South Africa's narrow transition may result in a multiracial authoritarian regime or at best an extremely limited democratic polity.[44] A coalition of the middle could order an authoritarian crackdown on opposition.[45]

Second, the parties of the center still have incentives during a transition to think about future elections, in which they will campaign independently. There will be strong temptations to undercut their partners in the transition to score points for a future election. If the parties in the middle succumb to this temptation, the net result will be continued ungovernability.

Even in the absence of an agreement about the transition, the National Party has been accused of such double-dealing. One newspaper editor, for instance, castigated NP behavior in the wake of Chris Hani's assassination

as narrow-minded electioneering at the expense of its new partner in negotiations, the ANC:

> White South Africans are being asked to contemplate a future as a pow-
> erless minority, under the ANC, yet when the ANC leaders try to deal
> with a national crisis which is beyond the power of government to con-
> trol, the white politicians snipe viciously from the sidelines, and bluster,
> and try to save themselves by attacking the ANC moderates.[46]

If we acknowledge that a coalition of the middle can just as likely lead
to an authoritarian condominium on power or ungovernability as to "enlightened" democratic decisions, then we must look elsewhere for sources
that can bring about the necessary enlightenment.

Adam and Moodley believe that rationality will ultimately lead to optimal choices by South Africa's leaders. Like Price, Adam and Moodley
stress that German-style social democracy and economic codetermination
is "the most rational and also the most likely scenario for South Africa."[47]

But Adam and Moodley err in seeing rationality as the key process
that will lead to the optimal solution for South Africa. The road to democracy in South Africa is strewn with traps in which limited self-interest can
lead rational actors to choose self-defeating courses of action. There are
plenty of strategic games left in the transition, where rationality in and of
itself will not provide the normatively best answer.

One such game of strategic interaction is mentioned above—both
sides will simultaneously try to create new institutions and campaign for
a future election. Such a situation can easily descend into Machiavellian
undercutting of one's erstwhile partner, and at some point the game can
become pure competition instead of cooperation. If players know that the
cooperative game will end, there is incentive to defect at the earliest point.

Another strategic trap lies in the creation of the institutions themselves. With the exception of the far right and the leaders of some of the
"independent homelands," South Africans see the benefit of new institutions. Almost any democratic dispensation will provide a pareto-superior
solution to the one that exists now. But pareto-superiority does not lead to
automatic compromise: the different solutions to the bargaining problem in
South Africa provide widely different distributional benefits to the competing parties.[48] As the parties jockey for distributional benefits, they risk
creating a situation where their eventual choice cannot hold because of ungovernability.

It is a leap of faith to believe that self-interest or rationality will lead
to enlightened policies and launch South Africa's democracy on a powerful positive basis. The only solution is one dismissed by Horowitz: cognitive change or "enlightened" self-interest, where groups "see the long-term
superiority of social contracts over simple contracts."[49] Only through

learning and problem solving will the parties of South Africa step beyond short-term limited self-interest.

Points of Optimism

We can take two examples straight from Horowitz's book that suggest learning and redefinition of self-interest have played a part so far in the creation of a new South Africa: the ANC's positions on proportional representation (PR) and federalism.

The choice of an electoral system in a new constitution will have distributional consequences. The rule governing representation—plurality or proportional—influences the number of seats in a legislature that political parties can win. For instance, with a plurality rule, a party with plurality support can win a much greater proportion of seats in a legislature than the proportion of votes it receives nationally in an election. Therefore, if a party believes that it has majority support, it will press for a plurality rule in the hopes of gaining disproportionately.

In the South African case, if we focus solely on distributional gains to be earned from the institutional rule governing representation, then the African National Congress would be best served by the plurality rule. Yet the ANC's formal constitutional proposals support proportional representation. Sisk, in his analysis of institutional choice in South Africa, suggests that the ANC has done so because it has been able to include a criterion in its decisionmaking of fairness over and above limited self-interest. Proportional representation will ensure that minority groups in South Africa are represented in parliament; it will educe cooperation by encouraging coalition building.[50]

Only through tortured reasoning can the ANC's institutional choice of PR be attributed to narrow self-interest. One such explanation, for example, states that the ANC's choice of proportional representation stems from its white SACP members, who desire to control the electoral process through the party list system.[51] The problem with this explanation is twofold. First, it shows that "self-interest" can be evoked post hoc to explain any choice—in this case the power of "self-interest" is maintained by changing the pursuer of self-interest from the ANC as a party to white SACP members in the ANC. Second, by changing the pursuer of self-interest to a small clique within the ANC, the explanation implies that the ANC as an organization and African non-SACP leaders in the ANC are either stupid and don't know their own self-interest or are completely powerless to fend off the machinations of the SACP cabal. Neither implication stands well with what we know of individuals such as Cyril Ramaphosa, Thabo Mbeki, Bridgette Mbandla, or other ANC leaders.

When we turn to the issue of federalism, we see a dramatic shift in ANC institutional choice. Writing just two years ago, Horowitz states baldly, "Consider the persuasion it will take to convince the extra-parliamentary opposition that federalism is not just a cleaned-up version of the homelands policy. The bias against federalism is so strong that, according to a sophisticated UDF leader who concedes its possibilities, it is not worth the immense effort it would take to change minds."[52]

But in 1993, the ANC negotiating committee conceded the desirability of strong regionalism (and in fact their proposals fit some Western versions of federalism). Could it be that the discussion, debate, and exchange encouraged by works such as Horowitz's, by seminars presenting U.S. and German versions of federalism could actually have impressed members of the ANC to see it as a good in and of itself? The elite survey that I cited earlier, found that in 1992 almost 20 percent of ANC leaders preferred a federal constitution as their first choice.[53]

The analysis of institutional choice relates to the key economic transformation necessary to sustain democracy in South Africa. Both Herbst and Price in this book agree that South Africa's economy needs to be overhauled. While Price believes that a strong central government, in conjunction with labor and business, can deliver the compromises necessary for South Africa to be economically competitive in the world, Herbst argues that the high-wage policy of the unions will pose an insurmountable barrier to job creation and economic growth.

From a narrow focus on organizational interest, Herbst is certainly correct. After all, why should we expect powerfully organized unions to forego higher wages—a policy that would be at odds with their raison d'être? From a long-term perspective, however, it is in the interest of the unions to have a thriving South African economy. If there was reciprocity from business in South Africa to reinvest profits to create jobs, then there would be an incentive for the unions to hold the line on wages. The key to realizing Price's hopeful scenario is through a process of debate and problem solving to get the main corporate actors in South Africa to redefine their limited definitions of self-interest.

Will such redefinition on economic policy take place in South Africa? Again there are hopeful signs. The unions have begun to think creatively about the need for job creation. They have expanded their concerns to include the needs of the unemployed. The mutual fund that the labor unions have established to invest worker pension funds places foremost among its priorities of social criteria for investment whether a company has an active policy and proven track record of job creation. The National Economic Forum (NEF) in South Africa has already forged the beginnings of a consensus among government, business, and labor on tough questions of strategy and sacrifice.

Conclusion

The South African case suggests that there is a dialectic at work in the creation of stable democracy. Changes in society, culture, and economy (*Transformation 1*) lead to a narrow transition to democracy. But in the course of creating a highly explosive situation that leads elites to a democratic compromise, the initial transformation also creates attitudes, rules, and conditions that work against democratic consolidation. The task of the elites who negotiate the narrow transition is then to work for *Transformation 2*—progress toward eliminating economic inequality, the creation of a culture of tolerance, and the building of a healthy civil society that will then support and sustain democracy. The alternative is for transition to remain in crisis, with the formal political rules of elites sorely out of touch with the informal rules and interactions of the society. The result can easily slip into authoritarianism or the collapse of the state.

In the South African case, the changes (and continuities) of the 1980s provide a mixed legacy for building a long-lived democracy. South Africa's leaders have the opportunity to pursue policies that can continue to transform society and build a firmer base for its democratic compromise. But let us be very clear: whether or not South Africa's leaders pursue such policies will not be merely a product of rationality or self-interest. Leaders rationally pursuing their self-interest in South Africa can just as likely lead to multiracial domination, a one-party state, or chaos. In the end it will be the ability and willingness of the actors to define their self-interest in new ways, to create together the need for a stable, democratic South Africa.

Will they do so? The question is not simply about optimism or pessimism. It is about what drives the calculations of leaders and whether learning, problem solving, and community will mitigate the excess striving of limited self-interest and the dictates of power.

Notes

1. Giuseppe DiPalma, *To Craft Democracies: An Essay on Democratic Transitions* (Berkeley: University of California Press, 1990).

2. Terry Lynn Karl, "A Research Perspective," in *Transitions to Democracy: Proceedings of a Workshop*, Commission on Behavioral and Social Sciences and Education, National Research Council (Washington, D.C.: National Academy Press, 1991), p. 32.

3. René Lemarchand, "Africa's Troubled Transitions," *Journal of Democracy* 3, no. 4 (October 1992), p. 101.

4. Terry Lynn Karl, "Dilemmas of Democratization in Latin America," in Dankwart Rustow and Kenneth P. Erikson, eds., *Comparative Political Dynamics: Global Research Perspectives* (New York: HarperCollins, 1991), p. 168. Emphasis mine.

5. Nancy Bermeo, "Shortcuts to Liberty," *Journal of Democracy* (Spring 1991), p. 116. See also the criticism of Adam and Moodley that the burgeoning literature on democratic transition ignores key economic components of distribution that make possible consolidation of democracy, in *The Negotiated Revolution: Society and Politics in Post-Apartheid South Africa* (Parklands, South Africa: Jonathan Ball, 1993), pp. 32, 224.

6. Robert Price, *The Apartheid State in Crisis: Political Transformation in South Africa: 1975–1990* (New York: Oxford University Press, 1991).

7. Ibid., Chapters 2, 5, 6.

8. The assertion that South Africa is likely to become a one-party state is found in Donald Horowitz, *A Democratic South Africa? Constitutional Engineering in a Divided Society* (Berkeley: University of California Press, 1991). The possibility of South Africa becoming an "elite cartel" that will rule through "multiracial domination" is found in Heribert Adam and Kogila Moodley, *The Negotiated Revolution: Society and Politics in Post-apartheid South Africa* (Parklands, South Africa: Jonathan Ball, 1993).

9. Jeremy Seekings, "From Boycotts to Voting? The Extra-parliamentary Opposition in Transition," *Die Suid-Afrikaan*, February–March 1993, pp. 33–34.

10. Such concerns were raised by Anthony Marx and Richard Tomlinson in their presentations at the SAIS Country Day Conference, April 13–14, 1992. The very issue of maintaining a healthy balance between the state and civil society has led one thoughtful policy analyst to recommend that U.S. foreign policy overwhelmingly provide direct assistance to NGOs and grassroots organizations in Africa, instead of channeling aid money through the state. See Michael Clough, *Free at Last? U.S. Foreign Policy Towards Africa in the Post Cold War Era* (New York: Council on Foreign Relations, 1992).

11. Seekings, "From Boycotts to Voting?" p. 34.

12. Khehla Shubane, "Civics as a Building Ground for a Democratic Civil Society," *Die Suid-Afrikaan*, February–March 1993, p. 36.

13. See Stephen Friedman, *The Elusive "Community": The Dynamics of Negotiated Urban Development* (Johannesburg: Centre for Policy Studies, 1993).

14. Ibid.

15. Horowitz, *A Democratic South Africa?* p. 89.

16. Adam and Moodley, *Negotiated Revolution,* p. 31.

17. Quote by Mandela in *Argus,* May 8.

18. Adam and Moodley, *Negotiated Revolution,* pp. 221–222.

19. Hennie Kotzé, *Transitional Politics in South Africa: An Attitude Survey of Opinion-Leaders*, Research Report No. 3, (Stellenbosch, South Africa: Centre for International and Comparative Politics, 1992), pp. 65–76.

20. Kate Manzo and Pat McGowan, "Afrikaner Fears and the Politics of Despair: Understanding Change in South Africa," *International Studies Quarterly* 36 (1992), pp. 1–24; and Hermann Giliomee, "Broedertwis: Intra-Afrikaner Conflicts in the Transition from Apartheid," *African Affairs* 91 (1992), pp. 339–364.

21. Manzo and McGowan, "Afrikaner Fears and the Politics of Despair," pp. 1–24.

22. Adam and Moodley, p. 222.

23. Frederik Van Zyl Slabbert, *The Quest for Democracy: South Africa in Transition* (London: Penguin, 1992), p. 91.

24. Timothy Sisk, *The Illusive Social Contract: Democraticization in South Africa* (Princeton: Princeton University Press, Forthcoming). To get an incisive and thorough examination of the development of the negotiating positions of the

major parties in South Africa and their potential for reaching a stable agreement, this manuscript is mandatory reading.

25. Stephen Lewis, *The Economics of Apartheid* (New York: Council on Foreign Relations, 1990), p. 136.

26. Bob Tucker and Bruce Scott, eds., *South Africa: Prospects for Successful Transition*, Nedcor/Old Mutual Scenarios (Cape Town: Juta, 1992), p. 27. The exercise defines successful transition as combining stable democracy, rising real incomes, reasonable distribution of incomes, and stable social fabric (p. 15). The definition combines the fact of transition—a change in the formal political rules—with what many see as prerequisites of sustaining such a transition. By doing so, it substitutes a definition—"stable democracy is the presence of a set of conditions"—for a hypothesis—"the presence of a set of conditions is likely to lead to stable democracy." Thus, the Tucker and Scott book categorizes Argentina, Bolivia, Brazil, Nicaragua, Eastern Europe, and the Soviet Union as "unsuccessful transitions to democracy" (p. 22), which would probably shock the peoples and analysts of these countries to know that their experiments have already failed! I have tried to put forward here that it is an open question whether democracy can survive as a set of formal institutions without progress on various transformational issues. This at least opens the possibility that transitions in difficult objective situations are not determined to fail. It also allows the possibility that there may be a significant lag time for narrow transitions to address key transformational issues. Normatively, it puts forth transformation as a needed policy to sustain transition and does not insist on transformation before transition.

27. Adam and Moodley, *Negotiated Revolution*, pp. 133–148.

28. See "'Outsiders Turn Daveyton into Killing Fields,'" *Weekly Mail and Guardian*, July 30–August 5, 1993; and "'His Sin Is that He Speaks Proper Zulu—in Jo'burg,'" *Weekly Mail*, June 18–June 24, 1993.

29. Donald Rothchild, "An Interactive Model for State-Ethnic Relations," in Francis M. Deng and I. William Zartman, eds., *Conflict Resolution in Africa* (Washington D.C.: Brookings Institution, 1991), p. 194; and Stephen John Stedman, *Peacemaking in Civil War: International Mediation in Zimbabwe, 1974–1980* (Boulder: Lynne Rienner, 1991), pp. 11–20.

30. Adam and Moodley, *The Negotiated Revolution*, p. 25.

31. Donald Rothchild, "An Interactive Model for State-Ethnic Relations," pp. 200–205.

32. Adam and Moodley, *Negotiated Revolution*, pp. 203–205, 159.

33. Ibid.

34. See "Taxpayers to Pay for State Blunders," *Weekend Argus*, February 13–14, 1993.

35. One analyst estimates that fraud-related crimes in South Africa over the last five years may involve upwards of $120 billion. See Njoroge Kariuki, "Economic Crimes in South Africa: A Hidden Legacy of Apartheid," *Southern Africa Political and Economic Monthly*, May 1993, pp. 41–42. See also *Financial Mail*, July 9, 1993, p. 42.

36. Adam and Moodley, *Negotiated Revolution*, p. 219.

37. Horowitz, *A Democratic South Africa?* p. 280.

38. Ibid., pp. 231–238 on crisis of distribution; pp. 205–214 on need for strong presidency.

39. Ibid., pp. 245–279.

40. Horowitz adopts a fairly narrow conception of learning. He dismisses the ability of individuals to learn from historical experiences or cases other than their own and, oddly enough, he makes no allowance for the participants learning from

works such as his own book. Yet over the last four years, the two main political parties in South Africa, the ANC and the National Party, have had teams of experts studying the global literature on constitutional engineering. People such as Horowitz, Arend Lijphart, and Roger Fisher have traveled to South Africa and advised the parties on optimal institutions. And the ANC's constitutional committee has travelled to Germany and the United States to study how federalism works in practice.

41. Ibid., pp. 271–279.

42. Ibid., p. 280 and p. 151.

43. Ibid., pp. 280–281.

44. Adam and Moodley, *Negotiated Revolution,* pp. 215–218.

45. Slabbert, *The Quest for Democracy,* pp. 80–83.

46. Ken Owen, "Where, at This Time of Crisis, Are Our Leaders?" *Sunday Times,* April 18, 1993.

47. Adam and Moodley, *Negotiated Revolution,* p. 210.

48. On the failure of pareto-superior solutions to explain institution creation, see Jack Knight, *Institutions and Social Conflict* (Cambridge: Cambridge University Press, 1992), pp. 34–37.

49. Horowitz, *A Democratic South Africa?* p. 151.

50. Sisk, *The Illusive Social Contract.*

51. R. W. Johnson, "PR at Work: A Case Study," *Die Suid-Afrikaan,* February–March 1993, pp. 20–21: "The proposed system is the product of an unholy alliance between the government on the one hand and the SACP leadership on the other—with African nationalism as the common fall guy . . . As the government stared aghast at the new electoral arithmetic, it naturally underwent a deathbed conversion to PR as the only way to moderate the tide of African nationalism. Ironically, the SACP elite, which occupied the commanding heights within the ANC, was coming to the same conclusion. The key white and Indian cadres of the SACP would have no chance of winning seats amongst their own communities under the old single member system. African nationalist sentiment would also make them extremely difficult to elect in black townships. How well would even a Joe Slovo or Mac Maharaj fare in a Soweto seat facing a PAC candidate preaching racial nationalism?" Ironically, an opinion poll of South African blacks showed that Joe Slovo was the second most popular South African leader, behind only Nelson Mandela. See *The Cape Times,* July 17, 1993.

52. Horowitz, *A Democratic South Africa?* p. 243.

53. Kotzé, *Transitional Politics,* p. 60.

3

South Africa: Economic Crises and Distributional Imperative

Jeffrey Herbst

The first postapartheid government will be faced with a dual economic challenge: to redistribute wealth and to fundamentally reform basic institutions and policies so as to permit faster growth. Indeed, long after the first nonracial election, unhappiness caused by lasting inequalities and perceived economic underperformance will be destabilizing, no matter how elegantly the new constitution is drawn. This chapter will first briefly review the economic situation facing the first postapartheid government. It will then sketch the political dynamics that will develop around two critical aspects of economic policy: fiscal balance and wage policy. These issues are critical to the health of the postapartheid economy; indeed, failure to be successful on both issues will torpedo all hopes of significantly increasing South Africa's growth rate. In addition, the two issues go to the heart of the constituency problems the ANC will face when making economic policy.

The South African Economy: Unequal and Inefficient

South Africa is classified by the World Bank as an upper-middle-income country with a per capita income of $2,530 in 1990, which makes it comparable to Mexico, Uruguay, and Venezuela.[1] As is now well known, the aggregate income figures hide significant inequalities.[2] Table 3.1 provides a summary of income share over time by racial group. While there has been some improvement in black income shares lately, the society is still obviously unequal. Indeed, 53 percent of Africans live below the poverty line compared to 2 percent of all whites; and to further these inequalities, the government spends R2,910 for each white child in primary school compared to R660 for each African child. There are 307,000 vacancies in white schools despite the fact that 2 million blacks are without school places; there are 180,000 whites in managerial, technical, and professional positions compared to only 3,000 blacks; there has been one black mine manager in seventy-seven years; and there is not a single black actuary.[3]

Table 3.1   Income Shares over Time

	1960	1970	1980	1990
African	19.9	19.8	24.9	33.0
Asian	2.1	2.4	3.0	3.9
Coloured	5.6	6.7	7.2	9.2
White	72.5	71.1	64.9	53.9

Sources: IMF, *Economic Policies for a New South Africa,* Occasional Paper No. 91 (Washington, D.C.: IMF, 1992), p. 4; and Max Frankel, Pollak Vinderine, et al., *Platform for Investment* (Johannesburg: Frankel, Max Pollak Vinderine, 1992), p. 72.

Note: These statistics come from a variety of sources that did not use the same methodology. However, the general trend reported in the table is supported by other surveys and anecdotal evidence.

What is less commonly known is that the South African economy has performed poorly since the mid-1970s and that structural bottlenecks in the economy will pose major problems for future leaders. Between 1946 and 1974, the South African economy grew at a real rate of 4.9 percent annually. However, over the next decade, growth dropped to 1.9 percent per year and for the decade of the 1980s, the economy expanded at a rate of only 1.5 percent.[4] In the 1980s, on balance, no new jobs were created in the nongovernment sector.[5]

A substantial portion of the economic crisis stems from a problem in garnering productive investment. Even in the period from 1963 to 1973, when South Africa managed to match the growth rate of the average Third World country, it was investing 25 percent of GDP compared to 15 percent for comparable countries.[6] Since 1973, gross fixed investment has declined markedly. Given that South Africa needs to invest 14 percent of GDP just to replace existing capital stock, the current figure of 16 percent suggests that overall fixed investment is barely increasing.[7] Part of the problem also stems from government investment that was manifestly unproductive. Much of government investment went into creating the sham structures of apartheid (e.g., capitals, airports, electrical stations for the "homelands" where all Africans not working directly in the white economy were supposed to be dumped) and strategic projects (notably the SASOL coal to oil program and the MOSSGAS natural gas to oil program) that were implemented to counter sanctions and that now have no economic rationale.

Productive investment was further stunted by the inward-looking industrial program that the South Africans have continued since 1948. As a result, while South Africa's traditional markets were growing by 6 percent annually between 1960 and 1990, its manufactured exports were increasing by only 2 percent.[8] Nongold exports did not increase relative to GDP because the South Africans believed that the gold industry was of utmost importance to the economy, a perception enhanced by sanctions.

The poor investment climate also deterred investors. The periodic political crises, including the Sharpeville massacre in 1960, the Soweto killings in 1976, and the township violence between 1984 and 1986 (which eventually led to the declaration of a nationwide state of emergency), each prompted large flights of capital.[9] International sanctions, the constant threat of unrest, and the all-pervading uncertainty of white life in South Africa further crimped the investment climate.

Finally, part of the South African investment failure in the 1980s can also be explained by a crisis of savings. In 1985, in the midst of the international debt crisis, banks led by Chase Manhattan reacted to the outbreak of violence in the townships by calling in South Africa's loans. Overnight, the country became a net exporter of capital. The forced repayment was achieved by running a significant current account surplus, which averaged 3 percent of GDP for 1985–1991 compared to a deficit of 2.5 percent of GDP for the period 1982–1984. Since the total level of savings remained roughly constant, the foreign debt (and the government deficit) was funded out of investment. Indeed, investment as a percentage of GDP declined from 26.4 in 1982–1984 to 20.2 in 1985–1991.[10]

South Africa faces a similar problem with human capital. Its failure to educate blacks caused the economy to begin to suffer from a severe shortage of skills in the 1970s. Currently, the ratio of managers to workers is one of the world's lowest, at 1:42 compared to 1:10 in developed countries.[11] Further, for every 100,000 people in each racial classification, there are 4,467 white students in tertiary education but only 665 Africans.[12] The country will probably continue to lose skilled workers through emigration.[13]

Therefore, any upswing in the economy will quickly cause a rapid rise in imports because of underinvestment in the manufacturing base. Indeed, Finance Minister Derek Keys calls the current account South Africa's "heart-lung machine" because even during the recession of 1989–1992 it was under pressure.[14] Further, sustained growth will also stoke inflationary pressures due to the shortage of skilled labor and other bottlenecks in the economy.[15] The Reserve Bank will be forced to respond to the deteriorating current account and inflation by clamping down on the money supply and thereby limiting growth. As a result, the successor government will probably experience a baseline economic growth rate in the first five years after the transition not substantially higher than the population growth rate of 2.7 percent.[16] Of course, exogenous shocks, especially caused by an increase in the price of oil or severe drought, could dampen even this mild performance.

The ANC's Economic Beliefs

Any new government would find the economic problems facing South Africa to be daunting. Indeed, many countries have found simply the task of removing structural inefficiencies and making the economy more outward-looking to be exceptionally difficult, without worrying about distributional implications. In South Africa, the unique analytic difficulties and uncertainties posed by the economic challenge are aggravated by the state of economic analysis in the African National Congress. The ANC was taken almost completely by surprise when it was legalized by de Klerk in 1990. Up to that point, it had devoted minimal attention to economics because it was focusing on the armed struggle and, as a broad front for those who opposed apartheid, it had little interest in detailing economic proposals that could only alienate potential constituencies. Indeed, the ANC has always been more like a church for all those who opposed apartheid than an actual political party with a defined ideology and program of action. The years in exile required nothing more. As a result, the extremely vague Freedom Charter adopted in 1955—which seemed to demand the nationalization of the mines, banks, and monopolies—was still considered the last word on the ANC's economic doctrine when it was legalized in 1990.

   In addition, while the ANC never believed it would shoot its way into power, the economic analysis that was done assumed that during the endgame the ANC would be dealing from a position of strength with a demoralized and divided government. The fact that de Klerk leaped ahead of the power curve and entered into negotiations as the confident head of a strong government forced the ANC fundamentally to review the assumptions behind many of its policies. As Mike Morris perceptively notes:

> Much of the left . . . inherited from the 1980s a tradition that shunned effective struggle within institutions. . . . These reflex social conceptions are wholly unsuited to the tasks that clutter the democratization process . . . this transition is about the inclusion of the opposition into existing institutions still inhabited by the same old social forces and people.[17]

Jeremy Cronin of the South African Communist Party is even more emphatic: "The ANC-led alliance has not developed an adequate strategy for
struggle in the post–February 1990 situation."[18]

The managed transition that South Africa is experiencing does provide
the ANC with significant advantages, notably an economy and infrastructure that are largely intact. However, taking full advantage of those benefits implies enormous restrictions on the Congress's freedom of action. In
particular, the Congress will have to pay far more attention to business
confidence and property rights in order to reassure the private sector than
it had anticipated. It is the gradual realization that an intact private sector
offers enormous benefits cum significant restrictions, which probably ex-
plains the ANC's quick abandonment of much of the rhetoric of nationalization that had featured so prominently during the decades of exile.

Finally, as the ANC was legalized, its ideological universe collapsed.
The commitment to socialism among ANC members varied tremendously,
but there is no doubt that the fall of communism left the organization without even a set of first principles. Of course, the collapse of communism
and de Klerk's legalization of the ANC are closely related. It was de
Klerk's tactical insight that the fall of the Berlin Wall provided a unique
window of opportunity for whites to negotiate a transition from a position
of strength. The speech of February 2, 1990, is extraordinary as a "state
of the nation" address because it discusses the opportunities posed by the
changed international environment before reviewing events in South
Africa.[19] Further, the message that much of the ANC's old economic beliefs were antediluvian has been amplified by the United States and by European countries, which have become much more important to the ANC.
The transition from "struggle economics" to a nuanced debate about the
country's future has thus been extremely difficult.

Macroeconomic Balance During the Transition

Unless South Africa's macroeconomy is in balance, there will be no hope
of implementing reforms to accelerate growth, and the country, given the
persistent underinvestment in the economy, will teeter on the brink of a
stabilization crisis. Unfortunately, the first postapartheid government will
inherit a difficult financial situation. Since 1975, general government expenditures increased faster than the growth rate of the economy irrespective of economic conditions. As a result, the government developed what
now appears to be a structural deficit. The deficit had not been severe until
1992/93, when it threatened to rise above 9 percent of GDP, a worrying
harbinger. Further, the composition of government spending is troublesome. Government spending on consumption rose from 18.5 percent of

GDP in the mid-1970s to 31 percent during the late 1980s. In contrast, the government's capital expenditure dropped from 7 to 4 percent of GDP during the same period.[20] Therefore, most of government borrowing is to cover consumption.

The fundamental problems with fiscal policy in South Africa were made clear by Finance Minister N. C. Havenga in 1950. He pointed to the "unstable climatic conditions of the country, the enormous financial burden involved in the advancement of the Black peoples, and the relatively small number of income taxpayers who have to bear this burden."[21] Indeed, throughout the 1980s, South African tax revenue became much more dependent on individual taxation as revenue from gold collapsed and company profits were squeezed by recession and high wage settlements. In 1991, direct taxation of individuals accounted for 56 percent of total direct taxation; in 1975, the figure had been only 25 percent.[22] South Africa's fiscal problems would have been evident earlier except for the brief gold boom in the early 1980s, which caused a temporary spike in revenue collection.

Not surprisingly, whites pay the overwhelming amount of income tax since they are disproportionately wealthy. In 1987, whites accounted for 72.1 percent of taxes, down only somewhat from 77 percent that they contributed in 1975. The tax burden of South African whites, at 32 percent of total income, is high by middle-income country standards and comparable to industrial country levels.[23] While the ANC says its wants to raise revenue by making the tax structure more progressive,[24] it is hard to see how this will generate significantly more revenue given the already high marginal rates that most South Africans now incur because of bracket creep.

In this context, the fiscal demands that a new government will face will be immense. Servaas van der Berg estimates that if other government expenditures (currently 17.9 percent of GDP) were held constant and just education, social pensions, health, and housing were increased to white levels, total government expenditures would increase from 27.4 percent of GDP to between 42.4 and 48.7 percent.[25] Such a level of increase is, of course, impossible. However, even to increase services somewhat is an extraordinarily expensive proposition, given the relative size of the impoverished portion of the population. For instance, simply to raise educational spending for Africans to the level currently received by Indians or coloureds would imply an increase in expenditures of between 65 and 109 percent, an impossible proposition given the relatively high amount that South Africa already spends on schooling.[26]

In addition, a future South African government will be faced with tremendous pressures to increase spending in other areas. For instance, the next government may want to create jobs programs for the millions unemployed, institute supplemental feeding programs for the 2.3 million people—including 1 million children—who are malnourished, begin programs to help the estimated 12 million people without adequate water supplies, and help electrify the homes of the 80 percent of all blacks without power.[27] Further, there will be demands to bring black farmers into the agricultural support system by providing them with, among other things, roads, seeds, fertilizer, outreach support, and depots. Finally, public sector employment is bound to increase significantly as the next government attempts to reward its own supporters and begins the long effort of making the civil service more representative.

There will be an "apartheid dividend," although it will not be significant enough to solve the fiscal problems of the next government. First, some government spending has already been redirected. Desmond Lachman and Kenneth Bercuson estimate that whites received 56 percent of total government benefits in 1975 but that this had declined to 35 percent in 1987.[28] The benefits whites receive from government today are even lower, especially given that much of the direct cost of operating schools has been passed on to white parents. There will also be savings from administrative duplication once the homelands are abolished, although van der Berg estimates these at only 1 to 2 percent of GDP.[29] Finally, there will be some significant savings from defense spending, probably in the range of 1.5 to 2 percent of GDP.[30] The best guess is that the short-run savings will amount to 2 to 4 percent of GDP.[31]

Of course, a new government is likely to find that government spending is "sticky" in the downward direction. First, transition rules, especially regarding civil service protection and pensions, will reduce the absolute amounts that can be saved. Since the National Party is guaranteed to have a prominent place in the new cabinet, side agreements on white perquisites can be expected to severely limit the apartheid dividend in the short term. Indeed, the NP will be especially concerned with the fate of white civil servants because, over the last forty-five years, government employment has been central to the economic upliftment of the Afrikaners who are still core voters for de Klerk's party. The fiscal implications of power sharing probably have not been fully analyzed by the ANC. Second, downsizing is an extraordinarily complicated task, which a new government will not have the administrative talent to complete quickly. Bureaucrats can also be expected to wage vicious guerrilla campaigns to protect their turf, with the result that actual savings will always be less than projected. Difficulties in producing short-term fiscal savings are especially important to note because the ANC is counting on extremely large efficiency gains to fund new programs.[32]

Finally, much of the apartheid savings will have to be invested, rather than consumed in the short term, if the country is to address its structural problems. Finance Minister Keys estimates that the percentage of the economy devoted to investment will have to increase from the present 16 percent to between 24 and 25 percent of GDP if a 3.5 percent growth rate is to be sustained.[33] If half of the increase in investment comes from reductions in government spending, perhaps an optimistic projection, the apartheid dividend will be all but eliminated.

It is unlikely that foreign savings will be able to ameliorate significantly a future government's fiscal problems. South Africa will be able to become a net borrower again on international capital markets, although it will have to be careful not to borrow too much too quickly. Debt servicing already accounts for 17 percent of total government expenditures, and this figure cannot be increased too rapidly or South Africa will fall into a debt trap.[34] There will also be an increase in foreign aid, although it too will not be enough to reduce pressure on the fiscus. Surprisingly, South Africa is already one of the world's most significant recipients of project aid.[35] Under a new government, many donors will simply shift this aid from the nongovernmental organizations that run the programs to public authorities. Additional foreign aid will be forthcoming but will be limited in a world that must also fund the reconstruction of the states of the former Soviet Union, Central Europe, and the rest of Africa. The Nedcor/Old Mutual estimates that overall capital inflows will go from zero in 1992 to 2 percent of GDP in 1995[36] seems reasonable but will not provide much budget relief given the cost of the programs listed above.

The result of the enormous potential demands on the fiscus plus the limited possibilities for raising revenue in the short term is a near guarantee of a destabilizing fiscal deficit. To their credit, leaders of the ANC seem to realize this problem. As Cyril Ramaphosa, general secretary of the Congress, noted, "Macroeconomic populist pitfalls which can have the opposite of good intentions in the medium to long-term have to be avoided."[37] Similarly, Tito Mboweni of the ANC's Department of Economic Policy has said that "something radical will have to be done to control the level of government expenditure."[38] The ANC is probably the only liberation movement in history to speak of financial discipline before it assumes power.

However, the means by which the ANC will actually limit expenditures is less clear. The ANC faces an enormous expectations crisis among blacks long discriminated against. The reasonable statements made in front of international bankers will inevitably give way to irresponsible populist promises during a hard-fought election.[39] Indeed, given that the ANC has had neither the time nor the analytic resources to address the enormously complicated questions it faces, the current conservative statements on fiscal policy should be seen as spin control as the ANC desperately searches for a policy appropriate to the situation it finds itself in. In particular, the Congress has raised its constituency on redistributive economics and has not even begun to develop a vocabulary to explain to its erstwhile supporters why they may have to wait before they see the fruits of the liberation struggle. Given that the current government was not able to resist the demands of 5 million relatively affluent whites, it is hard to see how the next government is going to be able to turn down the demands of 32 million mainly poor people who have suffered from institutionalized racism for decades.

Further, any rhetoric from the ANC on fiscal discipline will feed into the critique of radicals that the Congress is "selling out." Winnie Mandela, estranged wife of ANC leader Nelson Mandela, summed up the feelings of many when she charged in December 1992 that a deal was being negotiated between "the elite of the oppressed and the oppressors" and warned of "the looming disaster in this country which will result from the distortion of a noble goal in favor of a short-cut route to parliament by a handful of individuals." She elaborated on this critique in January 1993 when she claimed in a widely printed newspaper article that "the NP elite is getting into bed with the ANC to preserve its silken sheets and the leadership elite in the ANC is getting into bed with the NP to enjoy this new-found luxury."[40]

Finally, the nature of a significant portion of the ANC's constituency suggests that it will have a relatively short grace period before it is under extreme pressure to deliver the political goods. In most of Africa, it is true that governments survived for many years despite the fact that not only were they not providing significant benefits to their population but real incomes were declining.[41] Other African governments were able to avoid popular protest because 70 to 80 percent of their populations were atomistically dispersed throughout the rural areas and, as a result, could not organize easily. However, the South African population is urbanizing quickly. Between 1985 and 1990, the percentage of Africans in the cities rose from 45.5 to 50.8 percent. It is estimated that by the year 1995 more than 60 percent of the total population will be in the cities. Further, as opposed to most African countries at independence where the ruling party was the only organized political force, South Africa's population is highly politicized and there are already numerous politicians and groupings around which aggrieved urban residents can coalesce. In addition, many of the young people in particular have heard for almost their entire lives ANC demands that they make the country ungovernable until the demands of "the people" are met. There is every reason to believe that such rhetoric will come back to haunt the ANC.

A further aggravating factor is that, in contrast to most other African countries at independence, machine guns, mortars, and land mines are readily available in South Africa because of the long armed struggle, the regional arms markets that developed from the conflicts in Mozambique and Angola, and the greater local capacity to produce weapons. Thus, a future South African government will face a much more demanding population that is more concentrated, easier to organize, and better armed than was the case in the rest of the continent.

Once in power, the ANC will probably try to retain most of its constituency by widely distributing increases in government spending even at the cost of incurring a high deficit. The historical pattern of trying to gain maximum popular support is so deeply engrained in the ANC that it will have great difficulty shedding certain groups to avoid deficit spending. However, failure to control government expenditures will have damaging effects in the short and long terms. In the first few years after the transition, especially if the economy does begin to grow, the Congress will face a stabilization crisis. Damaging disinflationary policies will have to be adopted, which will result in lower growth rates. In addition, politicians will be less than anxious to remove South Africa's highly protectionist trade regime and liberalize import controls, reforms that must take place if the country is to become more export oriented.

In the long term, failure to control government spending, especially on consumption, will prevent the investment crisis from being addressed. The failure to increase investment will not be evident for several years; indeed, high government spending may make it appear that the economy is doing well. However, unless there is a massive redirection of spending from consumption to investment, South Africa will never be able to generate the resources it needs to grow and address the inequalities fostered by apartheid.

The only way for the ANC to address its structural fiscal problems will be to make some brutal calculations regarding its constituencies. In particular, the ANC will have to determine which groups it wants to support and which groups it can afford to neglect in order to bring the government budget into balance. Also, it will have to calculate which government programs to fund in a meaningful manner and which of the many neglected services in South Africa will have to remain so for a considerable period of time. The ANC's structure as a very broad church fronting for a host of different constituencies is simply inappropriate for the fiscal situation that it will confront.

The Labor Unions and the ANC

It is only natural that attention be devoted mainly to the inequalities between whites and blacks in South Africa. However, there are also severe economic and political inequalities within South Africa's black population that will help structure postapartheid politics and may have critical implications for the ANC in particular. Calculations for 1990 reveal that African wages in the (primarily urban) manufacturing sector were approximately R12,000 and in mining R9,200, while in the informal sector operators earned an average profit of R4,800 and their employees received incomes of approximately R2,700 annually. Worst off of those employed were laborers on farms who earned R2,400 and rural dwellers in the home-lands who received R1,400 each year.[42] It is estimated that 51.3 percent of the total working population, or 8.4 million people (almost all of them black), work in the informal sector or in subsistence agriculture.[43]

Second, the African work force has become more differentiated. Over the past fifteen years, there has been an increase in the number of profes-sional, semiprofessional, clerical, technical, and nonmanual skilled jobs outside of mining and agriculture.[44] While whites continued to dominate the skilled sector, Africans penetrated the semiskilled, technical, and non-manual positions (increasing from roughly 1 million to 2 million). Those Africans that had the skills to rise in the labor hierarchy as some of the discriminatory practices dropped away benefited, while unskilled blacks were faced with significant job losses.[45] Unemployment increased signifi-cantly and many more people were forced into the informal sector.

Finally, there are, even before the transition to nonracial rule, signifi-cant power asymmetries emerging within the black population, primarily due to the power of black unions. Following the Wiehan Report in 1979, the government recognized unions with African members. Probably the most important reason for this recognition was the need, constantly stated by business, to ensure orderly collective bargaining.[46] As businesses be-came more dependent on skilled and semiskilled African labor, the old form of industrial relations whereby managers issued diktats to a floating group of nonskilled workers (who often responded with wildcat strikes) became unprofitable. Businesses found that they had to have reliable means of communication with their increasingly skilled and permanent African workers. In turn, the legalization of the unions further aggravated the trend toward employing small numbers of skilled African workers while shedding a greater number of the unskilled.[47] The unions are partially responsible for the fact that real wages have increased in recent years despite the recession, poor productivity growth, and rapidly rising unemployment.[48]

Ironically, the African unions, spawned by business, became the most powerful domestic opposition to the white government in the 1980s. Un-like much of the rest of the opposition, they had no real fears of being de-clared illegal, had a well-defined and militant membership, and, due to for-eign funding, had durable administrative structures. Since 1990, while the ANC has floundered during the understandably difficult process of trans-forming itself from a guerrilla movement based in Lusaka to a political party based in South Africa, the unions have grown increasingly strong, led by the 1.2 million–member Congress of South African Trade Unions (COSATU).[49] Indeed, South Africa has lost more days to strikes in the last

five years than it did in the previous seventy-five.[50] Karl von Holdt is undoubtedly correct to claim that "it [COSATU] is the biggest and most
powerful organization in the democratic movement."[51]

Further, COSATU, along with the South African Communist Party, is
formally in alliance with the ANC. Ramaphosa, the day-to-day leader of the
ANC and the most likely successor to Mandela, was previously the head of
the black miners union, the most powerful labor group in the country. While
much can change quickly during the run-up to an election, it appears that
COSATU will continue its ties with the ANC during South Africa's first
universal election and therefore be in an exceedingly strong position to at
least influence labor- and wage-related issues.[52] However, COSATU and the
other unions have made it clear that, unlike many other labor groupings in
Africa, they will not join a future government and thereby become coopted.
They are confident that "the labour movement is powerful enough to block
any strategy of economic development that seeks to sideline or exclude or-
ganised workers."[53] COSATU has a place at the National Economic Forum,
and Mboweni of the ANC has said that he expects that "the role of trade
unions in economic policy formulation would increase in the future and
membership would continue to grow."[54]

The asymmetry within the African workforce between the small number who are unionized and workers (especially on farms) who are not
members of unions, who are unemployed, or who work in the informal or
subsistence sectors is striking.[55] To aid those blacks who are not formally
employed, South Africa should adopt a low-wage/high-employment policy. However, COSATU has specifically rejected the idea of its members
making sacrifices to increase employment. As Ebrahim Patel, COSATU's
representative at the National Economic Forum, argued, "A low wage
strategy is not possible because the labour movement is too strong and we
will resist it. Then, too, a low wage strategy will not work in a relatively
industrialized society."[56] As a result, especially in the context of unmet expectations due to fiscal limitations, a future government may be tempted to
appease the unions by granting wage increases to civil servants and favor-
ing organized labor more generally even if this means effectively discrim-
inating, especially against unorganized Africans in the informal sector of
townships and in the distant rural areas. Such a policy would be potentially
damaging to the overall economy's ability to create jobs and its international competitiveness. South African productivity rates have not been impressive, and while there will be some immediate gains once politically inspired strikes end, it would be quite easy for the country to become a
high-wage economy, if it is not one already.

The dangers of a high-wage economy at this point in South Africa's
development can be easily observed by comparing it to the newly industrializing countries (NICs) of East Asia. In these countries, wages were

held artificially low for decades because of government repression, lack of unionization, and a shared willingness to make sacrifices because, especially in South Korea and Taiwan, economic prosperity was tied to national security. The savings generated from not having to pay wages appropriate to the level of development were invested in the economy and eventually led to very high growth rates. South Africa aspires to the income levels that the NICs have achieved, but its high wages will prevent it from generating the necessary savings to fund investment.

The unemployed and those working in subsistence agriculture and the informal sector will have only limited options in the face of organized labor's political power. COSATU has made a series of largely rhetorical pledges that it seeks to ally itself with the unemployed and rural dwellers, but these claims can be dismissed rather easily.[57] The unions do not have the administrative structures for such alliances nor the political will, given the need to protect their own members during what will undoubtedly be difficult economic times to come. The ability of the ANC itself to be a counter to the unions by directly representing the interests of nonunionized workers and rural dwellers is also limited. Since they are already organized and relatively militant, workers will inevitably demand considerable attention from the ANC itself. Further, the ANC has had difficulty in establishing a political infrastructure in the countryside that would enable it to transmit demands from rural dwellers. In its 1986 conference at Kabwe, the Congress itself noted its relatively poor job in mobilizing Africans in the homelands.[58] Not surprisingly, organization in the rural areas has been an extremely difficult task since legalization in 1990, and it has naturally been much easier to establish party structures among supporters in the urban areas.

Many in South Africa, and some within the labor movement, have argued that the trade unions must adjust to the realities of the 1990s by becoming less militant. However, it is unclear if the leaders of the unions could deliver their membership if they agreed to a program that sacrificed wage demands to help improve business viability and create more jobs for those who are now economic outsiders. Forged in the crucible of the liberation struggle, the unions have become strong by constantly demanding more and by being fused commercial-political foci of opposition to "the white system." As South Africa moves through the transition, it is unclear if they can become more like traditional labor unions and compromise.

It is possible that, in the end, the need to create jobs will force a future ANC government to confront the trade unions and thereby shatter the alliance with COSATU. However, ANC leaders will be extremely reluctant to make such a break, because if labor were to form its own political grouping, it would undoubtedly take with it a significant portion of the Communist Party and many members of the ANC. Further, a public conflict with labor might threaten the entire constituency structure of the ANC. Once the ANC is in power and the glue of opposition to apartheid cannot hold its many different constituencies together, the appearance of a critical fault line with labor may encourage all kinds of other groups centered on ethnic, regional, or occupational identities to demand greater concessions or threaten to leave.

Conclusion

Since February 1990, the ANC has rethought many of the economic policies that it had previously advocated. However, it has not even begun to examine if its relationship with its varied constituencies still makes sense given the economic challenges facing the next South African government. Indeed, it would probably be unrealistic to expect the ANC to make tough decisions about who should be its core supporters and who might be sacrificed in light of tough economic realities until after the first election. However, at some point the ANC must make this difficult political calculation if it is to avoid a destabilizing fiscal crisis and the possibility of existing workers who are already relatively better off doing well at the expense of the much larger but unorganized group of Africans in the informal sector, on farms, and in the rural areas. It is likely that shedding some constituencies after decades of operating as a Great Coalition of all those who opposed apartheid will be even more traumatic for the ANC than the renunciation of long-cherished economic policies that has occurred since 1990. However, the ANC cannot afford to be all things to all people.

Notes

Research for this chapter was funded by a Fulbright scholarship. I am also grateful to Princeton University for its generous leave policy.

1. World Bank, *World Development Report 1992* (Washington, D.C.: World Bank, 1992), p. 219.
2. See generally Francis Wilson and Mamphela Ramphele, *Uprooting Poverty: The South African Challenge* (Cape Town: David Philip, 1989), p. 18.
3. Commonwealth Expert Group, *Beyond Apartheid: Human Resources in a New South Africa* (Cape Town: David Philip, 1991), pp. 7–8; Andrew Donaldson, "Financing Education," in Ronnie McGregor and Anne McGregor, eds., *McGregor's Education Alternatives* (Cape Town: Juta, 1992), p. 296.
4. Stephen Gelb, "South Africa's Economic Crisis," in Stephen Gelb, ed., *South Africa's Economic Crisis* (Cape Town: David Philip, 1991), p. 4.
5. Frankel, Max Pollak Vinderine, eds., *Platform for Investment* (Johannesburg: Frankel, Max Pollak Vinderine, 1992), p. 5.
6. Bob Tucker and Bruce R. Scott, *South Africa: Prospects for Successful Transition* (Cape Town: Juta, 1992), p. 70.

7. *Argus*, October 31, 1992, p. 22. Some evidence suggests that in 1993 fixed investment will drop below the replacement rate and, for the first time since 1947, South Africa's fixed capital stock will fall. *Financial Mail*, March 18, 1993, p. 35.

8. Tucker and Scott, *South Africa*, p. 80.

9. Brian Kahn, "Foreign Exchange Policy, Capital Flight, and Economic Growth," in Iraj Abedian and Barry Standish, eds., *Economic Growth in South Africa* (Cape Town: Oxford University Press, 1992), p. 88.

10. Brian Kahn, Abdel Senhadji, and Michael Walton, "South Africa: Macroeconomic Issues for Transition," Informal Discussion Papers on Aspects of the Economy of South Africa No. 2 (1992), pp. 4–6.

11. C. B. Strauss, "Education in South Africa," *Standard Bank Economic Review*, August 1990, p. 3.

12. These figures underestimate the problem because 22 percent of Africans in tertiary education are in teachers' colleges, compared to only 4 percent of white students. *SA Barometer* 6 (November 1992), p. 92.

13. One survey suggested that 250,000 young whites, out of a total white population of 5 million, are considering leaving the country. *Business Day*, January 11, 1993, p. 1.

14. Interview, Cape Town, February 26, 1993.

15. Standard Bank, "Preconditions for Structural Adjustment in South Africa," *Standard Bank Economic Review*, August 1992, p. 1.

16. Loots's estimate of 3 percent annually for the first five postapartheid years seems reasonable. Lieb J. Loots, "Budgeting for Postapartheid South Africa," in Glen Moss and Ingrid Obery, eds., *South Africa Review 6* (Braamfontein: Ravan Press, 1992), p. 471.

17. Mike Morris, "Who's In, Who's Out," *Work in Progress*, no. 87 (March 1993), p. 8.

18. Jeremy Cronin, "The Boat, the Tap and the Leipzig Way," *African Communist*, no. 130 (Third Quarter 1992), p. 41.

19. The de Klerk speech is reprinted in Robert Schrire, *Adapt or Die: The End of White Politics in South Africa* (New York: Ford Foundation, 1991).

20. Statistics in this paragraph are from Estian Calitz, "The Limits to Public Expenditures," in Graham Howe and Pieter le Roux, eds., *Transforming the Economy: Policy Options for South Africa* (Durban: Indicator South Africa, 1992), pp. 79–83.

21. Quoted in G. W. G. Browne, "Fifty Years of Public Finance," *South African Journal of Economics* 51 (March 1983), p. 147.

22. Calculated from South Africa, *Statistical Bulletin*, no. 8 (1990), p. 7. Taxes on the gold industry accounted for 19 percent of revenue in 1975 but only 2.7 percent in 1991.

23. Desmond Lachman and Kenneth Bercuson, *Economic Policies for a New South Africa*, IMF Occasional Paper No. 91 (Washington, D.C.: IMF, 1992), pp. 28–29.

24. ANC, "Discussion Document on Economic Policy," in *Economy: Growth and Redistribution* (Cape Town: IDASA, 1991), p. 11.

25. Servaas van der Berg, "Redirecting Government Expenditure," in Peter Moll, Nicoli Nattrass, and Lieb Loots, eds., *Redistribution: Can It Work in South Africa?* (Cape Town: David Philip, 1991), p. 79

26. Lachman and Bercuson, *Economic Policies*, pp. 22–23.

27. Poverty figures from Margie Keeton and Gavin Keeton, "Gearing Up for the Long Road: The Challenges of Poverty in South Africa," *Optima* 38 (Novem-

ber 1992), p. 129; and Francis Wilson, "Poverty, the State and Redistribution: Some Reflections," in Nicoli Nattrass and Elisabeth Ardington, eds., *The Political Economy of South Africa* (Cape Town: Oxford University Press, 1990), p. 235.

28. Lachman and Bercuson, *Economic Policies,* p. 28.

29. Van der Berg, "Redirecting Government Expenditure," p. 80.

30. Lachman and Bercuson, *Economic Policies,* p. 22.

31. Loots's estimate is 3 percent. See Loots, "Budgeting," p. 474.

32. Interview with Tito Mboweni, Johannesburg, February 18, 1993.

33. *Argus,* February 4, 1993, p. 17.

34. Ronnie Bethlehem, "Debt Trap Threatens Future Government's Ability to Function," *Business Day,* January 6, 1993, p. 4.

35. *Business Day,* January 6, 1993.

36. Tucker and Scott, *South Africa,* p. 165.

37. Quoted in *Argus,* October 10, 1992, p. 4.

38. "To Get Out of the Mess Won't Be Easy," *Mayibuye,* December 1992, p. 22.

39. The ANC estimates that it will receive roughly 50 percent of the vote based on 68 percent of all Africans' votes, 20 percent of coloureds' votes, 30 percent of Indians', and 3 percent of whites'. However, there is a high degree of uncertainty concerning turnout, especially among Africans. Frankel, Max Pollak Vinderine, *Platform for Investment,* p. 44.

40. *Sunday Argus,* January 24, 1993, p. 18.

41. I have examined the dynamics of political survival in the face of economic failure in *The Politics of Reform in Ghana, 1982–1991* (Berkeley: University of California Press, 1992).

42. Peter G. Moll, "Conclusion: What Redistributes and What Doesn't," in Moll, Nattrass, and Loots, *Redistribution,* p. 119.

43. Up from 32.6 percent of the population (2.3 million) in 1960. D. J. Viljoen, *Labour and Employment in Southern Africa: A Regional Profile, 1980–1990* (Johannesburg: Development Bank of Southern Africa, 1990), p. ES1. Of course, estimates for employment in the informal and subsistence agriculture sectors, as well as the number unemployed, are little more than guesses because of difficulties in measurement and the failure of government statistical services to gather information on the black labor force.

44. The number of skilled workers increased while the total number of jobs decreased due to technology, increases in wages, and a desire by employers to escape labor problems by using more machines.

45. Doug Hindson, "The Restructuring of Labour Markets in South Africa: 1970s and 1980s," in Gelb, *South Africa's Economic Crisis,* pp. 230–234.

46. See Commission of Inquiry into Labour Legislation, *Report,* reprinted in *The Complete Wiehahn Report* (Johannesburg: Lex Patria, 1982), p. 35.

47. Hindson, "Labour Markets in South Africa," p. 234.

48. South African Reserve Bank, *Annual Report 1992* (Pretoria: South Africa Reserve Bank, 1992), p. 19.

49. Total union membership stands at 3 million people. Union membership has grown by 11 percent a year since 1980. Andrew Levy & Associates, *Annual Report on Labour Relations in South Africa, 1992–1993* (Rivonia: Andrew Levy & Associates, 1992), p. 3.

50. Duncan Innes, "Labour Relations in the 1990's," in Duncan Innes, Matthew Kentridge, and Helene Perold, eds., *Power and Profit: Politics, Labour, and Business in South Africa* (Cape Town: Oxford University Press, 1992), p. 184.

51. Karl von Holdt, "What Is the Future of Labour?" *South African Labour Bulletin* 16, no. 8 (November–December 1992), p. 36.

52. See generally Jeremy Baskin, *Striking Back: A History of COSATU* (Johannesburg: Ravan Press, 1991), p. 434.

53. Von Holdt, "Future of Labour?" p. 36.

54. Quoted in *Business Day*, January 27, 1993.

55. Estimate for union membership is from Andrew Levy & Associates, p. 3, and Jeremy Cronin, "Is Nelson Mandela for Real?" *Work in Progress,* no. 87 (March 1993), p. 16.

56. Quoted in "Going in with Confidence," *South African Labour Bulletin* 17, no. 1 (January–February, 1993), p. 27.

57. Indeed, COSATU's initial efforts to organize the unemployed to prevent them from being used as scab labor failed badly. Baskin, *Striking Back,* p. 453.

58. See Jeffrey Herbst, "Prospects for Revolution in South Africa," *Political Science Quarterly* 103 (Winter 1988–1989), pp. 665–686.

4

At War with the Future?
Black South African Youth
in the 1990s

Colin Bundy

Youth Activism and Youth Crisis

When Nelson Mandela addressed a tumultuous crowd in Cape Town on the evening of his release from prison, he listed individuals, organizations, and groups that had sustained the struggle against apartheid. "I pay tribute," he said at one point, "to the endless heroes of youth. You, the young lions. You, the young lions, have energized our struggle."

Mandela's salute to the young lions—and the roar with which it was received—acknowledged a central and persistent feature of South African politics over the past twenty years: the prominence of student and youth activism. It echoed the recognition and rhetorical convention, during the mid-1980s, that "the youth" constituted not only a generational but a political category. "Young people," observed a black newspaper in April 1986, "have experienced an unprecedented moral ascendancy. They are known universally as 'the youth,' the legion of black teenagers who for the last two years have provided the shock troops of a nationwide popular insurrection. This has been a children's war."[1]

The hour of "the youth" had struck a full decade before this comment was made. The Black Consciousness movement of the early 1970s was based mainly in colleges and schools; the year-long youth revolt that began in Soweto in June 1976 not only qualitatively transformed youth activism but also cracked the mold of South African politics. Student protest became a mass phenomenon; school boycotts "became ordinary features of the political landscape in the towns and ghettoes: student organizations remained permanent fixtures at the forefront of political agitation."[2]

The spate of school boycotts in 1980–1981 was part of a wider process, a key element in the emergence of radicalized, localized community-based protest in black townships. In 1979, the Congress of South African Students (COSAS) was founded, and it became the most influential student activist organization that South Africa has known. COSAS broke with Black Consciousness, adopted the Freedom Charter, and became one of the largest and most dynamic affiliates of the United Democratic Front. It explicitly sought to mobilize school students on the basis of popular front opposition to apartheid, and it was the driving force in the early 1980s behind the proliferation of youth congresses. These included some students and some young workers but also members of an embittered and volatile army of young unemployed; they "infused a deeper, sometimes more desperate militancy into student and community politics."[3]

In 1984, COSAS spearheaded protracted school boycotts and initiated a mass stay-away in the Transvaal, together with trade unions and civic associations. For two days, 300 schools were deserted by 400,000 students, and over half a million workers downed tools. The stay-away was assessed as opening "a new phase in the history of protest against apartheid"[4]—a phase of united, disciplined, and strategically planned mass action. But this phase was being overtaken by another when that judgment was penned. Already in September 1984, townships in the southern Transvaal had erupted; and for the next two years the politics of organized joint demonstrations was largely replaced by a turbulent politics of direct action and quasi-insurrectionary confrontation.

Throughout the 1984–1986 period—until successive states of emergency imposed a shaky peace at gunpoint—student and youth activism delivered a considerable voltage to a highly charged circuit of resistance. Many members of student and youth congresses became full-time activists. High school students were initially involved mainly in school boycotts, but their energies were diversified in three main directions: participation in mass campaigns, service in alternative structures, and participation in direct action and street combat.

Student and youth activism in this period displayed a characteristic dualism. Intertwined with its strengths (intensity, energy, and creativity) were its shortcomings. Almost by definition, it lacked experience and sophistication; it tended to triumphalism or immediatism, anticipating imminent victory with little patience or strategic reality. Thus, the period saw black youths attend and run "alternative education" classes, build "people's parks" on waste sites, patrol the streets in anticrime campaigns, and form the front line in "defense committees." It also saw action based on a profound misreading of the balance of forces (typified in the boycotters' call for "liberation before education"); it witnessed how rapidly youth politics could pass from intensity to zealotry. "People's courts" dominated by youngsters veered from justice to vengeance; stay-aways and boycotts were sometimes enforced by intimidation; the term *com-tsotsi* accurately reflected social reality as lumpen elements became caught up in, or made cynical use of, inflammable social movements.[5]

These aspects of youth activism began to win the attention of social scientists during the 1980s. Black youth politics stamped itself onto the academic agenda. In the last couple of years, however, a different set of concerns has begun to emerge. It is voiced in various ways but has become audible not only in the media and among academics, but also within black communities, within political movements, and in student and youth organizations themselves.

The ruling class press and media often refer to a "lost generation," frequently and falsely explaining the crisis in black schooling as a consequence of the boycotts of the mid-1980s. In a similar vein, alarmist references are made to "Khmer Rouge teenagers" and "agents of anarchy." In a different register, left-wing journalists (writing in the educational supplement to the weekly *New Nation*) judge that black schools have been turned "into sites of despair and demoralisation," and that in student politics "on the ground active participation has collapsed." Senior officeholders in student organizations concurred. "The black student community is currently largely apolitical, demoralised and individualised," said Sipho Maseko, president of the Azanian Students Convention (AZASCO). The "real crisis" faced by students had created "a strong sense of demoralisation and frustration over the past two years," added Moeti Mpuru, national projects officer of the South African Students Congress (SASCO).[6] Political organizations are concerned about their ability to hold the loyalty of their youthful supporters—and there is mounting evidence that suggests that these anxieties are merited.

The focus of attention, in short, has shifted to different aspects of a multifaceted "youth crisis." Members of "the youth" are variously portrayed as the victims or vectors of that crisis. There is growing awareness of the desperate state of black schooling, steepling rates of unemployment and crime, disintegrating family structures, and the threat of AIDS. These various symptoms are sometimes collated in a malaise described, somewhat inadequately, as "marginalized youth." This was the title of an important conference held in June 1991 under the auspices of the Joint Enrichment Project (JEP).[7]

The JEP conference established overwhelmingly the severity, urgency, and complexity of the problems confronting young people in South Africa. The working group set up at its close has tried to ensure that the issues discussed at the conference remain on the political agenda of bodies like the African National Congress. It held follow-up workshops and sponsored a program of research; it created a network of regional youth development committees; and in June 1992, a National Youth Development Coordinating Committee was elected by delegates from over eighty youth organizations. A second national JEP conference on youth (the National Youth Development Conference) was held in March 1993.

Results of a major survey—the first systematic attempt to quantify marginalized youth—were presented at the 1993 JEP gathering. The category comprised young people, aged between sixteen and thirty, who "see themselves as having little or no future, who are alienated from their families or job or school, who are out of touch with, or hostile to, the changes taking place in South Africa, who have been victims of abuse and/or violence, who have a poor self-image, or who are not involved in any organisation or structure."[8] The survey sample suggests that 2,900,000 people (of whom 2,500,000 are Africans) are already marginalized from political, economic, and social processes.

The JEP initiatives have identified the need for research into the size, location, nature, and regional incidence of marginalized youth and have already made important contributions toward those tasks. This chapter does not undertake such research; rather, it is an attempt to suggest something of the context in which those details must ultimately be grounded.

The category "youth" has tended to be constructed in two ways in South Africa in recent years: Jeremy Seekings has usefully dubbed these "liberatory" and "apocalyptic" stereotypes.[9] The first refers to militant, politically active young people: the Young Lions or, simply, the comrades.[10] This usage tended to assume, implicitly or explictly, that the majority of young people evinced similar political characteristics.[11] Second, the term "youth" has been used as if it were more or less synonymous with what the JEP conference called "marginalized youth," characterizing young blacks as undisciplined, destructive, and dangerous. Both uses are ideological. The former defines youth according to a set of positive political characteristics; the latter essentially views youth as the vector of a compendium of social ills and anxieties.

This chapter does not focus upon either of these versions of what constitutes "the youth." It is concerned, rather, with drawing attention more broadly to the demographic and social salience of young people in South Africa at present, and to argue that there are specific pressures and problems that affect them. Kumi Naidoo has complained, correctly, that much of the work on youth produced in the 1980s "implied that youth constituted some kind of monolithic bloc"; and he warned that "failure to disaggregate the category led to unproductive generalisations."[12] Alas, much of what follows is open to similar objections, but that is the nature of the exercise. The following section identifies certain structural or objective factors that shape the experience of being young and black in South Africa today; the section "Violence and Youth Culture" sketches some of the symptomatic or subjective expressions of that experience; and the final section is a freehand sketch of some policy implications.

Structural Pressures

Demographics

Some absolutely basic pressures are demographic in character. South Africa is in a phase of accelerated population growth, with an annual increase of about 2.5 percent a year—which means that the total doubles in roughly thirty years (see Table 4.1). The rate of increase is far more rapid in the black population than in the white. In 1980, the fertility rate of African women was more than double that of white women.[13]

Table 4.1 Population Statistics and Projections

Year	Total population
1950	12,671,000
1960	15,841,000
1985	33,621,000
2000	47,000,000
2020	80,000,000

Source: Based on census figures and projections in South African Institute of Race Relations (SAIRR) surveys.

It is not only the size of the population but also its shape that is significant. The demographic profile is typical of a period of rapid growth in that a very large proportion of the population is youthful. South Africa, in fact, has "an extremely youthful population by world standards."[14] In South Africa today, half the population is under the age of twenty-one. In 1990, 42 percent of Africans were in the age bracket 0–14 years; and fully two-thirds of the African population were twenty-seven or younger. This phenomenon is especially pronounced where poverty is most acute—in rural areas and in the Bantustans.

There are also a number of demographic indications that many children are being raised in family units under considerable pressure. There is "a very high rate of family breakdown" among the African population (and, among white South Africans, "one of the highest divorce rates in the world").[15] There are extremely high rates of teenage pregnancy, of illegitimacy, and of female-headed households.[16] Surveys among young Africans show that teenage pregnancy is the most frequently cited problem.

Political considerations entirely aside, these demographic realities guarantee that in the foreseeable future there will be very large numbers of children and young adults experiencing a wide range of wants and deprivations. They ensure a continuing enormous pressure on educational resources. They make highly likely the persistence of youth unrest, and they underline the importance of the directions taken by youth culture. S. Burman is surely correct in her conclusion: "At present, any examination of current political developments and influences in South Africa must include children as an active and visible force." Under any government, the probability is that schools will continue to be both "laboratories and fortresses of resistance, providing raw political education to the pupils passing through them."[17]

Economics

The second and equally obvious set of structural pressures on young people is economic. In the 1970s, South Africa entered a period of economic recession. In the 1980s, crucially, South African capitalism did not share in the recovery evinced by the most industrialized and many of the newly industrializing economies. This is confirmed by a glance at the figures in Table 4.2. The all-too-familiar symptoms of slowed growth, falling investment, rising unemployment, and chronic inflation persisted—compounded by an international credit squeeze and balance of payments and exchange rate pressures.

Table 4.2   Annual Growth Rate of Gross Domestic Product

Period	Average Annual Growth (%)
1960–1969	5.9
1970–1979	3.1
1980–1989	1.3

Source: B. Godsell and J. Buys, "Growth and Poverty," in Robert Schrire, ed., *Wealth or Poverty? Critical Choices for South Africa* (Cape Town: Oxford University Press, 1993), p. 640.

The economic factor with the most direct implications for youth marginalization is unemployment. Unemployment statistics in South Africa are notoriously imprecise, but a recent considered estimate is that formal unemployment has risen from about 22 percent in 1970 to 30 percent in 1980 to 40 percent by 1989.

While the total of people out of work expands, the number of available jobs shrinks. Government statistics confirm that between 1984 and

1987 there was a *fall* of 200,000 jobs in mining, construction, manufacturing, electricity, transport, and the postal services. In Durban, heart of the second largest industrial region in the country, the labor force contracted by 20 percent between 1982 and 1984. In October 1991, the Development Bank of South Africa announced staggering figures: over the previous five years, only 12.5 percent of nearly 400,000 people entering the job market had found employment. The comparable rate in the early 1960s was 89 percent; in the early 1970s it was 49 percent.[18]

The great majority of these new job seekers are, of course, school leavers. Unemployment is a mass phenomenon in South Africa but is visited with particular ferocity on the age bracket 16–30. It is estimated that up to 90 percent of the unemployed may be under the age of thirty and that 95 percent of first-time work seekers in the 1990s will be under the age of twenty-five. A financial journalist concluded: "So a more accurate name for the unemployment crisis is the youth unemployment crisis."[19] (Findings published in March 1993 paint a picture less bleak in some respects. The Educational Policy Unit at the University of the Witwatersrand traced some 1,200 individuals who had matriculated in 1984 and in 1988. Over two-thirds of the 1984 class were either in wage employment or studying; for the 1988 school leavers, only 36 percent of the males and 23 percent of the females had found work. Crucially, however, this study dealt only with the small minority of African school children who completed matriculation [the certificate obtained on successful completion of the final year of high school]. In 1984, only 9 percent of children who had enrolled in primary school had stayed within the system until the end of high school.)[20]

The crisis in black schooling has been long in the making. It is rooted most directly in the system known as Bantu Education, introduced in 1955—although educational segregation is much older than that. The Bantu Education Act of 1953 allowed for direct state control over African education, and its framers specified that the content of such education should be in accordance with the subordinate status of Africans in South African society. H. F. Verwoerd (then minister of Bantu affairs, and later prime minister), piloting the bill through parliament forty years ago, was brutally frank:

> Native education should be controlled in such a way that it should be in accord with the policy of the state. . . . If the native in South Africa today . . . is being taught to expect that he will live his adult life under a policy of equal rights, he is making a big mistake. . . . There is no place for him in the European community above the level of certain forms of labour.

Bantu Education provides for rigid educational segregation. African children are taught in separate schools, with textbooks and syllabi designed for use in those schools, by teachers trained in separate institutions. Although spending on black education has increased substantially in recent years, it still lags well behind per capita expenditure on white children and has done little to remedy the physical backlog. High schools in African townships and in Bantustans have all the features one would expect from an educational system starved of funds: dilapidated buildings, crowded classrooms, inadequate equipment, high student:teacher ratios, and underqualified teachers. There is a dismayingly high dropout rate. Only 5 percent of African children who enter the school system make it to matriculation. Sixty percent do not complete primary school. Twenty-five percent of all dropouts are Grade I pupils.

An already dismal situation was transformed in scale during the 1970s and 1980s. For the first time, high school education for Africans became a mass phenomenon, the problems were multiplied, and pressure on existing resources grew worse (see Table 4.3).

Table 4.3   The Expansion of African Education in South Africa

Year	Total at School	High School	Matriculation
1950	779,534	23,316	449
1960	1,506,034	48,600	717
1970	2,741,087	122,489	2,938
1975	3,486,261	318,565	9,009
1980	3,532,233	577,584	31,071
1984	5,547,467	1,000,249	86,873
1987	6,644,859	1,474,333	157,274
1991	8,100,510	2,208,480	280,918

Source: Based on SAIRR surveys.

The education crisis deepened during the mid-1980s and has taken on new dimensions in the last few years. Politicized black school students demonstrated their rejection of the system by boycotting their schools in the 1980s; about 30 percent of all African schools were boycotted and the government closed about 90 percent of these for varying periods. Hundreds of thousands of scholars simply dropped out of school. Many who stayed or returned received patchy, disrupted, deteriorating schooling. Increasingly, there is reference among educationists of how the "culture of learning" is being eroded in African schools. "Simply put, this means that students and teachers alike have lost all faith in the system."[21]

The education crisis manifests itself in different symptoms and at various levels of intensity in different regions. Some vivid insights of how thoroughly destabilized black schooling has become in some areas are provided in a survey of 244 matriculation (final year) African high school students in nine Natal and KwaZulu schools. Most of the students had entered high school in 1985 or 1986—at the time, that is, "when the violence in Natal assumed war proportions." Actual violence, or the threat of violence, permeates the school experience. It affects travel to and from school, forces the unscheduled closure of schools, leads to the collapse of commitment and motivation on the part of the students, and permits the increasing penetration of the school yard and classrooms by criminal gangs.[22]

Urbanization

The demographic, economic, and educational features just outlined will be increasingly concentrated in the cities. This may seem a truism, but in the context of apartheid history it is extremely significant. Central to apartheid was a system of controls over labor, mobility, and residential rights. This was intended to keep the permanently urbanized black proletariat as small as possible, forcibly compelled huge numbers of people to reside in the Bantustans, and perpetuated the system of oscillating, low-wage migrant labor. In the mid-1980s, the machinery of urban influx controls and pass laws was abruptly scrapped. Economic decay and popular defiance of the restrictions exacted this important policy reversal.

In the 1990s, the collapse of influx control and the pressure of poverty in rural areas makes pell-mell urbanization inevitable. People will pour into cities even while unemployment rises and resources shrink. Much of the new urbanization will take the shape of "informal settlement" or squatting. Already, an estimated 7 million people live in squatter communities up and down the country. The demands exerted by this process for housing, health care, education, food, and employment will be intense. A very high proportion of the new urban population will be youthful. Specifically urban aspects of youth culture—like criminal gangs and street children—will reach new levels. One of the central facts of rapid urbanization is that it will no longer be possible, as it was under high apartheid, to export unemployed youth to the Bantustans and to marginalize them geographically.

Societal Violence

Perhaps the South African phenomenon most widely reported in the international media over the last decade has been mounting political violence. The most cursory recital of some of its forms conveys how familiar they became: guerrilla attacks; actions by police and army troops against protestors; the rise of local vigilantes; destabilization tactics throughout the region; the civil war in Natal and its extension to the townships of the

Witwatersrand, with the horrors of mass attacks upon squatter camps and migrant hostels; carnage on trains and buses; and the grim logic of vengeance and retribution.

Yet any discussion of violence that concentrated solely on these forms or that isolated the last ten years as the field of study would be seriously deficient. As an important study of the "epidemiology and culture of violence" reminds us, South African society is permeated by violence. Political violence "is only a small part of the overall picture," write Shula Marks and Neil Andersson, and must be seen as "but one of the many forms and varieties of endemic violence." Apartheid made industrialization in South Africa a particularly violent and traumatic process. The explicit repression of recent years

> comes on top of other forms of violence which are no less devastating and pernicious. The counterpart of the state violence is the continued violence of unnecessary death, disease, degradation and disability imposed by the racially structured inequalities and humiliations of the apartheid system. . . . It is as ubiquitous as the imperious attitude adopted each day by whites towards blacks, in the streets and shops, factories and farms of South Africa; and it is as quantifiable as the statistics of violent death and social dislocation.[23]

South Africa has extremely high rates of judicial executions, alcohol and drug addiction, motor vehicle and industrial accidents, suicide, homicide, and family murders. Between October 1989 and February 1991, 3,200 people died in "unrest" killings; barely remarked in the same period were 26,300 recorded murders. South Africa has the highest known incidence of rape in the world (2 million women have been raped since 1984; statistically, one of every two women is likely to be raped in her lifetime). The common factor between criminal violence and political violence is that black youths "feature prominently as both the perpetrators and victims of the violence."[24]

It is against this somber backdrop that political violence became more visible and more intense in the 1980s, with mass-based political protest and heightened state repression its most obvious indices. Young people played a central part in the waves of protest that ebbed and flowed between 1976 and 1989. Collated official figures show that for the period from 1984 to 1986, 300 children were killed, 1,000 wounded, 11,000 detained without trial, 18,000 arrested on protest charges, and 173,000 held awaiting trial in police cells. "To these grim figures should be added those of tens of thousands of children who have experienced deaths in the family . . . detentions of family members, being on the run from authorities, and numerous forms of intimidation."[25] The toll of deaths and injuries through political violence has increased since February 1990. Before then, it was primarily an overt conflict between the regime and the forces of liberation. Now, political violence takes a number of tangled and intractable forms between various antagonists with a range of motives. The most distinctive feature of contemporary political violence is that for the most part it takes the form of attacks by impoverished working-class Africans upon other impoverished working-class Africans. In part, this stems from the intensification of political competition between the ANC and Inkatha; in part, it stems from die-hard elements within the security and intelligence forces who have an intrinsic interest in internecine black violence. The following section tries to assess some of the consequences of the intensification of political and criminal violence in the 1980s and early 1990s.

Violence and Youth Culture

A number of commentators have posited as a direct consequence of intensified violence in the 1980s the "brutalization" of young South Africans, or their "internalization" of violence. The central point made is that violence generates violence, that violence is a learned response. Levels of "socialized violence" among children and youth appear to have increased during the 1980s as parental, school, and other institutional controls were weakened or swept aside. Society became simultaneously more violent and more polarized; and political behavior echoed this. Fierce activism was accompanied by shallow understanding. Its practitioners often inhabited a political landscape shaped by simplistic absolutes and perverted morality, where people were killed because they wore the wrong color T-shirt or walked on the wrong side of a boundary road.

Children who witnessed or initiated acts of violence developed a range of psychological and psychosomatic disorders. The psychologist Saths Cooper noted that there was "very little normality" in the lives of politicized children. There was, he added, "a very short gap between being the victim of brutality and beginning to commit it one's self. Distinctions between right and wrong, killing and not killing, attacking and defending, become blurred. . . . Personalities begin to fracture."[26] In October 1986, the South African Institute of Clinical Psychology warned of a generation of maladjusted children.[27] Even very young children internalized violence. Games played by toddlers involved Casspirs, AK-47s, ambushes, and funerals. Frank Chikane wrote in 1986:

> Life in the townships seems to have changed irrevocably. A township resident said, "When my two-year old daughter sees a military vehicle passing, she looks for a stone." Nursery school children are no exception. They too have learned the language of *siyanyova* [we will destroy].[28]

Recently, these emphases—analytical and anecdotal—have been challenged by Gill Straker. Her work is based upon lengthy interviews with sixty youthful activists from Leandra township who fled their homes in 1986 and whose subsequent experiences were explored in 1989. She distinguishes between "leaders" (the most resilient individuals in the group); "followers" (who did not become antisocial or dysfunctional as a result of their exposure to violence); and "casualties" (about 20 percent of the sample, who showed clinical signs of depression and anxiety). She finds, in sum, "little evidence of the Leandra youth having become a brutalised generation in the sense that their capacity for empathy and guilt . . . had been impaired." Equally, there was little to indicate that they would engage in indiscriminate violence, "although there was much to indicate that violence would be their response to particular sets of circumstances."[29] Similarly, an attitudinal survey focusing on leisure activities and preferences of "ordinary urban and peri-urban youth" also emphasizes that many youth "still express a positive attitude and remain receptive to opportunities" and rejects the "lost generation" stereotype.[30]

The most detailed attempt to gauge subjective responses to violence and unemployment suggested that a quarter of all young South Africans were "fully engaged with society" and needed no interventions on their behalf; that 43 percent of the sample were "at risk," showing some signs of alienation; that over a quarter (27 percent) were already marginalized; and that 5 percent (or just over half a million young people) were "lost"—that is, ejected by society, with no hope for the future and frequently involved in criminality and/or violence.[31] The same report proposed three "syndromes" of marginalization. *Antagonists* are "more politically alienated, racially antagonistic and hostile to the older generation"; *outsiders,* typically from broken families, are alienated from school or the work environment or are unemployed, have a poor self-image, and are often found in squatter settlements; *victims* have suffered abuse or have had wide exposure to political violence.[32]

Each of these studies warns against blanket generalizations, and they provide ways of differentiating young people and their needs. At the same time, certain sorts of violent behavior involving mainly young people appear to be increasing. The intensification of criminal youth gangs has been widely noted—graphically by Glenn Frankel in the *Washington Post,* who wrote that many youngsters "are coming of age into a society that offers no jobs and no hope." One result, he continues,

is a new generation of gangs like Soweto's "Zebra Force," whose leaders have boasted that they plan to rape every woman in the township under 26. They can be seen on the dusty streets of Soweto—angry young men in Nelson Mandela T-shirts. They move in packs, some brandishing weapons, stalking their own communities with the same passion their predecessors used to expend on battling white soldiers and police.[33]

Another symptom in 1992 was the extent to which violence had spilled over into schoolyards and classrooms. South Africa's school year begins at the end of January. The *Weekly Mail* reported in February 1992 that

> pupil power has taken over as frustrated youths turn on their teachers in the Transvaal townships. The Department of Education said this week it could not guarantee the safety of white teachers after a lecturer was set alight in his classroom. The incident followed a week of violence in which white staff were threatened by armed youths . . . inspectors were chased from the premises and black teachers were taken hostage. Thousands of pupils refused entrance to overcrowded schools are roaming the streets—while, in some, pupils have taken control of admissions. . . . Each year the crisis deepens as failure rates soar. Pupils have lost all faith in the system. Teachers, the most immediate representatives of authority, are often the butt of their anger.[34]

A year later, further school protests in Soweto were characterized by *City Press* as schoolchildren "on the rampage," and the writer deplored the "common knowledge" that in township schools children "call[ed] the shots."[35] Indeed, during 1993, not only the black media but also political leaders grew increasingly edgy about the disruptive capacity of marginalized youth. Peter Mokaba, president of the ANC Youth League, warned that youth, once in the forefront of the struggle, "have become spectators, Codesa watchers, and are disillusioned by the political process."[36] Further ground for concern by political organizations is provided by the survey carried out for JEP, which indicated that the proportions of "lost" and "marginalized" youth are appreciably higher in political organizations than in the sample as a whole.[37]

One of the most eloquent editorial pronouncements on the topic, which captured the realities of commitment *and* its costs, was written by Percy Qoboza, editor of *City Press,* in April 1986.

> If it is true that a people's wealth is its children, then South Africa is bitterly, tragically poor. If it is true that a nation's future is its children, we have no future, and deserve none. . . . [We] are a nation at war with its future. . . . For we have turned our children into a generation of fighters, battle-hardened soldiers who will never know the carefree joy of childhood. What we are witnessing is the growth of a generation which has the courage to reject the cowardice of its parents . . . to fight for what should be theirs, by right of birth. There is a dark, terrible beauty in that courage. It is also a source of great pride—pride that we, who have lived under apartheid, can produce children who refuse to do so. But it is also

a source of great shame . . . that [this] is our heritage to our children: The
knowledge of how to die, and how to kill.[38]

These chilling words were written at the height of repression, during the
state of emergency. South Africans have grounds for hope now that were
not available then. It is easier now to think in the future tense than it was
then. But the process of negotiations and the prospects of transition should
not be allowed to occlude the realities of brutalization among the youth.
Nor should expectations of a new political dispensation obscure for a moment the grave social and structural pressures that threaten further to alienate and marginalize huge numbers of young people in the years ahead.

Policy Implications

The question of what policy responses are appropriate and possible demands a chapter of its own. Here, very briefly, it may be useful simply to
identify some of the more obvious areas that must engage policymakers
and to pose the issue as a political challenge.

First, there is an acute need for imaginative and effective welfare programs that deal directly with some of the effects of youth marginalization.
It will be necessary to target such programs accurately: to distinguish, for
example, between the very different needs of school dropouts with no literacy or numeracy; those who completed primary schooling only and possess rudimentary educational skills; and those who dropped out during
high school or failed to pass the matriculation examinations.

Second, an area of priority is the transformation of South Africa's educational system. It is not merely that more resources will have to be directed to education, nor that the present proliferation of racially distinct
education systems must be replaced by a unitary, nonracial school system.
These are necessary but not sufficient responses to the crisis. There is also
the massive problem of inherited backlog, especially in the level and nature of teacher training; there are unresolved debates as to the balance between academic and technical training needed for the twenty-first century;
and at every step there is the question of how to ensure equality of access
to schooling in a society so profoundly unequal as South Africa is—and
will continue to be, even after the abolition of statutory apartheid.

The other area that virtually defines itself as a priority is job creation.
Attention is being paid to ways in which community organizations, small
employers, large employers, trade unions, funding agencies, and foreign
aid can play their part. But structural unemployment is of such a scale that
analysts are increasingly looking at examples of job creation on a mass
scale by governments. There has been a rediscovery of some of the New

Deal experiments of the 1930s, especially the Works Progress Administration, and advocacy of programs that would simultaneously create employment and build schools, roads, and other infrastructural requirements. Strikingly, this approach has been supported, on the one hand, by advisers to institutions of finance capital and, on the other hand, by left-wing researchers attached to the black trade unions.[39]

Perhaps the main thrust of the JEP conference in March 1993 was to call for a nationally driven youth development policy. The conference endorsed the creation of a National Youth Development Forum (NYDF), a broadly based body intended to "ensure the integration of youth as a key interest group" in national planning. The conference called on the new NYDF to implement a two-pronged strategy. It called for a specific program of interventions to address problems confronted by youth in fields like education, health, employment, and so on; but it also envisaged a long-term formulation of policies necessary to enable and empower youth in the longer term.

Frequently, however, such policy proposals are pitched in a technocratic and oddly apolitical register. Calls for "training" and "job creation" or for combating drugs, gangsterism, and violence often fail to consider the identity, capacity, and consciousness of young people themselves. They tend to be seen as policies that will be implemented on behalf of schoolchildren, school leavers, and young adults. They do not ask instead how such groups can be empowered, how youngsters can develop confidence and a sense of self-worth, nor how the energies and dynamism of youth can be harnessed to programs of rebuilding. (The JEP strategic proposals are largely—but not entirely—exempt from this criticism.)

In particular, the political question needs to be put. What lead might be given by politically active youngsters? To ask this is not to fall into the error of assuming that all or most of those in their teens or early twenties will be politically active. Seekings and others have punctured this myth, by showing that a large proportion of young black people are politically inactive or incurious, and animated by a wide range of other concerns and interests.[40] Even so, by comparative standards high levels of dynamism, participation, and social purpose were achieved in community-based organizations in South Africa in the 1980s. Large sectors of the youth were politicized. Similar levels of "mobilization for reconstruction" will have to be achieved in the 1990s if political programs addressing the problems confronting young people are to have a chance of succeeding.

For many young black South Africans—like those Straker interviewed in Leandra—the politics of negotiations has been a disillusioning experience. They feel sidelined, marginalized, and demobilized. They feel "that their contribution at the forefront of the struggle was being undervalued, that they were not sufficiently recognised as partners in the negotiating process."[41] The ANC Youth League has not recruited members in anything like the numbers that joined the youth congresses in the late 1980s; for some the formation of the League was "a dull, slow process, which has failed to capture the mood and imagination of the youth."[42] The articulate and engaged spokespersons for the National Youth Development Coordinating Committee warn that youth should be active participants in policy-making and negotiations and not be "further marginalized by being converted into tokens and rubber-stamps."[43] The challenge confronting the ANC and other political organizations should be seen as one of remobilizing its youthful constituency and of turning "spectators, Codesa-watchers" into activists.

When postapartheid elections are held, fully half of the projected electorate of 21 million will be voters under the age of thirty. About one-third of the potential voters will be under the age of twenty-five. How successfully the new government contends with the daunting agenda awaiting it will depend in large measure on how successfully it recognizes and responds to the problems that confront young South Africans on a mass scale. Young people must be taken seriously—which means enrolling them for change and not merely controlling them—if South Africa is to become a nation at peace with its future.

Notes

This chapter is a revised and expanded version of my "Introduction" in D. Everatt and E. Sisulu, eds., *Black Youth in Crisis: Facing the Future* (Johannesburg: Ravan Press, 1992), pp. 1–9.

1. *City Press,* April 20, 1986, quoted in S. Johnson, "'The Soldiers of Luthuli': Youth in the Politics of Resistance in South Africa," in S. Johnson, ed., *South Africa: No Turning Back* (London: Macmillan, 1988), p. 94.

2. M. Murray, *South Africa: Time of Agony, Time of Destiny* (London: Verso, 1987), p. 246.

3. L. Chisholm, "From Revolt to a Search for Alternatives," *Work in Progress* 42 (1986), pp. 14–19.

4. Labour Monitoring Group, "Report: The November Stayaway," *South African Labour Bulletin* 10, no. 6 (May 1985), pp. 74–100.

5. The phrase *com-tsotsi* combines the widely used appellation "comrade" with *tsotsi,* or "youthful gangster." A sensitive study of how youthful intolerance and violence affected a "people's court" is W. Scharf and B. Ngcokoto, "Images of Justice in the People's Courts of Cape Town, 1985–7: From Prefigurative Justice to Populist Violence," in N. C. Manganyi and A. du Toit, eds., *Political Violence and the Struggle in South Africa* (London: Macmillan, 1990), pp. 341–372.

6. Between October and December 1991, "Learning Nation" ran a six-part series on educational issues. The quotations here are from "Education Issues 3: Results and Prospects," *New Nation,* November 8–14, 1991; and "Education Issues 6: Organisation Then and Now," *New Nation,* November 29–December 5, 1991.

7. The JEP was established in 1986 by the Southern African Council of Churches and the South African Catholic Bishops Conference as part of a church initiative "to address the problems of young people roaming the streets as a result of the dislocation of the normal school process and lack of employment prospects." Some of the papers presented at the JEP conference are collected in Everatt and Sisulu, *Black Youth in Crisis.*

8. *'Growing up Tough': A National Survey of South African Youth,* designed and analyzed for the JEP by D. Everatt and M. Orkin of the Community Agency for Social Enquiry (CASE) (presented at National Youth Development Conference, Broederstroom, March 1993). The survey is based on a sample of 2,200 people, interviewed for up to one and a quarter hours each.

9. J. Seekings, *Heroes or Villains? Youth Politics in the 1980s* (Johannesburg: Ravan Press, 1993), pp. 2–10.

10. For a sensitive account of the ideological and cultural matrix of the comrades, see A. Sitas, "The Making of the Comrades Movement in Natal 1985–91," *Journal of Southern African Studies* 18, no. 3 (September 1992), pp. 629–641.

11. For example, G. Straker, *Faces in the Revolution: The Psychological Effects of Violence on Township Youth in South Africa* (Cape Town: David Philip, 1992), p. 19: "Among the youth it was the majority who participated in the eruptions [of 1984–1986]. The youngster who did not participate . . . was the exception rather than the rule."

12. K. Naidoo, "The Politics of Youth Resistance in the 1980s: The Dilemmas of a Differentiated Durban," *Journal of Southern African Studies,* 18, no. 1 (1992), pp. 144–145. See also Seekings, *Heroes or Villains?* pp. 1–2, 99–101.

13. C. Simkins, "Household Composition and Structure in South Africa," in S. Burman and P. Reynolds, eds., *Growing Up in a Divided Society: The Context of Childhood in South Africa* (Evanston, Ill: Northwestern University Press, 1986), pp. 28–29.

14. *Population Trends,* published for the Urban Foundation (Johannesburg, 1990), p. 13.

15. S. Burman and R. Fuchs, "When Families Split: Custody and Divorce in South Africa," in Burman and Reynolds, *Growing Up in a Divided Society,* pp. 119, 116.

16. Simkins, "Household Composition and Structure," pp. 32, 36–37; E. Preston-Whyte and J. Louw, "The End of Childhood: An Anthropological Vignette," in Burman and Reynolds, *Growing Up in a Divided Society,* pp. 361–362.

17. S. Burman, "The Contexts of Childhood in South Africa: An Introduction," in Burman and Reynolds, *Growing Up in a Divided Society,* p. 4; Johnson, "Soldiers of Luthuli," p. 96.

18. A. A. Ligthelm and L. Kritzinger-Van Niekerk, "Unemployment: The Role of the Public Sector in Increasing the Labour Absorption Capacity of the South African Economy," *Development Southern Africa* 7, no. 4 (November 1990); quoted in R. Riordan, "Marginilised Youth and Unemployment," in Everatt and Sisulu, *Black Youth in Crisis,* pp. 75–76.

19. R. Rumney, "Bring Back Roosevelt to Give the Jobless Jobs," *Weekly Mail,* January 17–23, 1992, p. 17.

20. These findings were reported in "All Is Not Lost . . . for Some Youths," in *Review Education,* supplement to *Weekly Mail,* March 12–18, 1993.

21. *New Nation,* "Education Issues 3: Results and Prospects," November 8–14, 1991.

22. Blade Nzimande and Sandile Thusi, *Children of War: The Impact of Political Violence on Schooling in Natal* (Education Policy Unit, University of Natal,

1991). The quotation is from p. 26. For a slightly earlier but similar study, see John Gultig and M. Hart, "'The World Is Full of Blood': Youth, Schooling and Conflict in Pietermaritzburg, 1987–1989," *Perspectives in Education* 11 (1990).

23. Shula Marks and Neil Andersson, "The Epidemiology and Culture of Violence," in Manganyi and du Toit, *Political Violence,* pp. 30, 32.

24. Steve Mokwena, "Living on the Wrong Side of the Law: Marginalisation, Youth and Violence," in Everatt and Sisulu, *Black Youth in Crisis,* pp. 30–31.

25. L. Swartz and Ann Levett, "Political Oppression and Children in South Africa," in Manganyi and du Toit, *Political Violence,* p. 265.

26. Quoted in Johnson, "Soldiers of Luthuli," p. 120. Cf. A. Thomas, "Violence and Child Detainees," in B. McKendrick and W. Hoffman, *People and Violence in South Africa* (Cape Town: Oxford University Press, 1990), p. 460: "Children, brutalized by physical and institutionalized violence have often, as a means of survival, resorted to violence themselves. In this process they are in danger of losing all sense of right and wrong."

27. Quoted in Straker, *Faces in the Revolution,* p. 2.

28. F. Chikane, "Children in Turmoil: The Effects of Unrest on Township Children," in Burman and Reynolds, *Growing Up in a Divided Society,* p. 343.

29. Straker, *Faces in the Revolution,* p. 105.

30. V. Moller, R. Richards, and T. Mthembu, "In Search of a New Morality," *Indicator South Africa* 9, no. 1 (Summer 1991), pp. 65–67; V. Moller, *Lost Generation Found: Black Youth at Leisure, Indicator South Africa Issue Focus* (Durban: Centre for Social and Development Studies, University of Natal, 1991), esp. pp. 6, 54–55, 57.

31. *Growing Up Tough,* survey for JEP conference, pp. 34–35. The degree of marginalization was calculated on the basis of a series of indices or "dimensions of concern": these included experience of abuse; family disintegration; gang membership and recidivism; attitudes toward the future and other races; self-esteem, etc.

32. Ibid., p. 39.

33. "Murderous Interregnum in South Africa," *Guardian Weekly,* week ending February 2, 1992. And cf. Mokwena, "Living on the Wrong Side of the Law," pp. 39–41.

34. *Weekly Mail,* January 31–February 6, 1992.

35. Quoted in the *Cape Times,* February 24, 1993.

36. *Cape Times,* February 26, 1993.

37. *Growing Up Tough,* survey for 1993 JEP conference, p. 38.

38. Quoted in Johnson, "Soldiers of Luthuli," p. 130.

39. Rumney, "Bring Back Roosevelt."

40. Seekings, *Heroes or Villains?* pp. 16–19, 98–99.

41. Straker, *Faces in the Revolution,* p. 134.

42. Seekings, *Heroes or Villains?* pp. 89–90.

43. "Presentation by the National Youth Development Coordinating Committee to the [JEP] National Youth Development Conference," March 1993, p. 9.

5

Local Negotiations
in South Africa

I. William Zartman

B eneath the headline-catching maneuvering of the South African parties through national-level negotiations and regime change lie myriad
local, limited, and lesser negotiations that lead, pace, contest, and support
the larger process.[1] Local negotiations are little stitches holding together
small pieces of the larger fabric of civil society. They may head in the
same direction as the larger regime-change negotiations, producing a har-
monious and supportive political structure; or they may run at cross-pur-
poses, in many directions, leaving snags and gaps and structures of inco-
herence. The two levels interact with and affect each other, struggling for
harmony and dominance. The pace and nature of national negotiations
have constantly influenced the outcome of local negotiations; the reverse
effect is in the longer run.

Beyond the fate of South Africa itself, local negotiations pose important
theoretical issues. One set of issues enters the state-society debate, at a crucial point in a nation's history. The relation between the two levels of negotiation has a powerful role in determining the nature of state-society relations
and indeed of the state itself. In South Africa, as in other states newly achieving control of their destinies, society is taking over state and defining it, but
not in a vacuum. Like society, the state already exists, a heritage from the
previous regime, and it too is fighting to impose itself on its society. This
does not refer to the white struggle for dominance over blacks; that is a separate struggle, between parts of society (or societies) over the control of the
state. Rather, there is both a level and a pattern of behaviors that are seeking
to impose themselves on society. Local negotiations pit local representatives
of society—essentially black citizens' and workers' groups—against local
representatives of the state—usually provincial authorities; but they also pit

the national protostate level of society—mainly the African National Congress—against the local representatives—mainly the civic organizations (civics). Thus, the local and national branches of society have their own competing agendas, fighting over autonomy versus control, respectively, as well as over the specific issue at hand. In previous notable cases—soviets in the Russian Revolution, comités de salut publique in the French—the state won. One may hypothesize the same outcome in South Africa, but the game is not up and the struggle itself has important consequences, whatever the result.

Another set of issues revolves around the concept of power. Local negotiations in South Africa in the early 1990s are asymmetrical, and therefore run against the basic assumption of equality underlying either the nature of the negotiation process or at least its successful occurrence.[2] How do negotiations take place under conditions of extreme asymmetry? What is the relation between negotiation and power generation? An intuitive hypothesis might be that negotiations await power generation, since power is needed to negotiate and equal power is an optimum condition. But might the process of negotiation itself also create power?

The subject of grassroots political organization and empowerment is one that has attracted the attention of both academics and activists. It responds to the deepest attachments of democratic ideals, where the people themselves practice their own governance, unimpeded by the conflicting interests and personal vulnerabilities of representative government. Aristotle's praxis and Rousseau's paradoxes centered around direct democracy, without resolving many of its problems. However, when the additional complication of a revolutionary situation is added, where local democracy faces an alien state—as in anticolonial movements, racial and neighborhood revolts, and classical revolutions—the drama of unequal power adds a new dimension to the debate over democracy.

This study is based on research conducted in Johannesburg, Pretoria, Pietermaritzburg, Durban, Port Elizabeth, Grahamstown, Cape Town, and their environs throughout July and August 1990 and revised in August 1992.[3] It aggregates the findings from analysis of some two dozen local negotiations, about which the following questions were asked: What is the power relation of the parties? What efforts are made by the weaker to overcome asymmetry? What is the relation between negotiation and the generation of power? What is the relation between the local negotiations and national negotiations on the same subject area, or national negotiations in general? What is the impact of local negotiations on the evolving nature of state and society?

Cases of Local Negotiations

Local negotiations can be divided into four categories: community (housing and services), labor, education, and security. It is significant that it was hard to find negotiations over education; its time had not yet come. There were also fewer negotiations—especially successful ones—in the security field, although, while intercommunal strife has worsened, more cases—including successes—have occurred since the National Peace Accord of September 15, 1991. It should be emphasized that not all negotiations have been successful; the best that can be said about some of them is that they have not ended. Still, even these contain lessons about the general practice and patterns of local negotiations. Not all of the cases studied are reported below, but only those that bring some broader insights or are typical of a larger number. The vignettes presented here are necessarily inconclusive and ongoing, not completed experiments but illustrative examples.

Community

Among community negotiations, the best-known and most significant is negotiation on the *Soweto rent dispute.*[4] Beginning in 1986, inhabitants of Soweto (South West township, near Johannesburg) refused to pay rent for inadequate services and housing that they had used for decades and were prohibited from owning. It took two years for the possibility of negotiation between the Soweto People's Delegation (SPD) and the white state's Transvaal Provincial Authority (TPA) to emerge, during which time the SPD built its organization and support and the two sides tested their strength. Evictions were blocked by organized crowds, confiscated furniture was replaced, and further occupation was barred. Any drastic retaliation by the TPA, such as service cutoffs or mass evictions, would provoke a reaction so massive as to render it impossible. The clincher came in October 1988, when the popularly rejected elections for the township council brought in the Sofasonke Party (SP), a populist organization that supported the boycott; the point is not that the SP was an alternative to the SPD but rather that even the official attempt at an alternative supported the boycott.

In April 1989, the consulting firm PlanAct suggested a five-point platform to the SPD for its negotiations with the TPA: (1) write off service and rent arrears, (2) transfer housing stock to residents, (3) upgrade services, (4) assess an affordable service charge, and (5) create a single tax base combining the white cities and their dependent black townships. SPD also insisted on a number of other far-reaching principles: (a) electricity supplies must be integrated, so as to adhere to the single tax base requirement; (b) the Soweto Township Council must be excluded from service determinations; (c) a new electric utility must be established, with community participation; and (d) fundings for services should come as grants from national, regional, city, or private sources, not as loans from the TPA.

At the same time, in a useful diversion, Eskom, the state electrical utility company, proposed an independent electrical utility for Soweto to be run by the township; the talks collapsed, largely over the relations between the local talks and national authorities, but in the process, the SPD learned a lot about the electricity business. On October 2, 1989, the SPD and TPA resumed negotiations on the basis of the five points; talks were again interrupted throughout January by the SPD to protest the existence of a separate development committee within TPA that made decisions for the body. When talks resumed, on January 27, the TPA had run out of money, and the government, refusing any further bridging aid, issued a July 31 deadline for the end of nonpayments, later extended by a month by President de Klerk. Over the summer, the TPA tried for some interim agreement, which the ANC vetoed. Finally, at the end of the deadline period, the two parties asked for a three-week period to reach an agreement with their parent organizations, which was achieved in principle in September 1990 with the acceptance of the five points. In June 1993, the decade-long rent boycott ended and Soweto was integrated under the Johannesburg and Roodepoort city councils, opening it to all city services and eliminating the township councils. The SPD was succeeded by a monitoring forum composed of all black political parties.

Although the initial agreement was only to study the implications of the five principles, it had constitutional significance. As a result of the SPD rejection of the township council, the central government repealed the Black Local Authorities Act of 1982, whereas the creation of new service utilities with consumer participation became an agenda item demand for national constitutional negotiations. Since 1990, negotiations have broadened to the Johannesburg metropolitan chamber, where the municipality, the SPD, the TPA, and others work out the allocation of services for the greater Johannesburg metropolitan area; while the results do not have the force of law, they are closely followed by the government and will be enacted when adopted by the chamber. By 1992, at least one metropolitan area, Kimberley in the Transvaal, had already held its own elections for an all-municipal council.

The negotiations themselves also had power significance. The SPD had been able to build up its local support and technical competence, impose unbearable costs on the TPA, and win some side contests along the way. The state was caught between two cost crunches—the high cost of providing services without payments and the threatened cost of civil unrest in the event of service cutoffs—and was a prisoner of its commitment to negotiate. That gave SPD the power to refuse to agree, and so keep the issue alive for purposes of mobilization. The civic association had also been able to block or coopt all other alternatives to itself and its program at a time when the government needed to show its willingness to resolve problems by negotiation, and it has consolidated and demonstrated its popular support and organizational ability. All of these elements provided power to the civic.

A much lesser case involved the *Salt River community hall* in Cape Town. The health and amenities committee of the Cape Town City Council (CTCC) planned to close the hall in 1988 because it was losing money; community organizations opposed the move. Some forty civics were involved, in various states of organization and rivalry, many of which would not join in the same meeting with others. A further complication that arose in many cases was a difference in political culture and experience between the two sides. The city council was an expeditious and businesslike body, controlled by the Democratic Party, whereas the civics needed time to confirm their organization, elect their spokespersons, develop a consensus, legitimize their positions, and overcome their sense of suspicion and inferiority vis-à-vis the city council. Following conciliation by a CTCC community relations specialist, the CTCC halted its plans and sought a meeting with the civics, which was finally achieved in July 1990, two years later. The civics challenged the council's reasons for closing, showed a need and use for the hall, established a need for upgrading, and made proposals. A joint working group of civics and the council was established to plan upgrading and then the establishment of a day-care center; in the process, extravagant demands of the civics were confronted with budgetary realities and pared down to realizable programs.

The result was not only a compromise of positions based on common interests, but also an education process for both parties. The city council learned to consult with affected populations through their ad hoc civic organizations; the community learned to organize and to present its views. In the process, both learned to develop a negotiation culture that gave legitimacy and understanding to the negotiations going on at the national level.

Although the Salt River dispute involved a relatively minor and specific issue, the process by which the authority agreed to consult the affected population and the civics agreed to meet with the authority was one that gradually came to characterize other larger issues, such as the decisions of the South African Committee on Sports on the use of the Greenpoint Stadium in 1989 or the response of the CTCC to the Thornhill Commission report on local governance in 1990.

At *Hout Bay* squatters sites, south of Cape Town, nearby white residents were disturbed in 1990 by the encroaching settlements of 2,000–3,000 squatters. In a new pattern of events, the Cape Provincial Authority (CPA) convened a meeting of both sides, in the presence of the Urban Foundation and, in exchange for a promise to freeze further influx, promised limited sites and services elsewhere, combining state land and private funds in a new approach. Again, the difference in styles and situations was used tactically: to a quick plea to accept the offer because of local residents' pressures, the squatters responded with an agreement to a lengthy study of the proposal because of other settlements' pressures against setting a harmful precedent. Because the state wanted to avoid a festering and eventually conflictual situation that would endanger private properties, the squatters' strength was found in their opponents' weakness. The result is close to "normal politics" or "petition politics," in which citizens petition their government for redress of grievances. The shadow of the future also hangs over the evolving pattern, since it gives impetus to national negotiations—to "normal politics" on the national level—while at the same time setting precedents for a new, presumably ANC government that will be hard to bear.

In *Zwela Themba,* the township of 25,000 near Worcester (100,000) in the western Cape Province, representatives of the township and appointed administrators from the city met regularly in 1990 in the presence of lawyers and city planners called in by the squatters. The township's Committee of 7 refused to cooperate in the meetings (which it had called for in the first place) until services were provided for the squatter settlements. It further demanded agreement on the principle of a single jurisdiction for the combined city and township, and rejected a development plan for Zwela Themba because it was treated separately from Worcester, although it accepted partial development measures.

On the side of the coloured community in Worcester, two competing civics had arisen, one ANC-dominated and one autonomous. When the community sought to join the township in support of the proposal, it was told to unite its civics first. The Cape Town–based Center on Inter-Group Studies of Hendrick van der Merwe proposed its services and helped consolidate the coloured group, then set up workshops for the three racial groups. The process was then in turn handed over to the Institute for a Democratic Alternative in South Africa of Frederik Van Zyl Slabbert for further workshops in democratic elections, in view of electing an all-municipal council. In mid-1992, however, the civics halted the process for fear that further progress would undermine the legitimacy of civic organizations in other municipal areas that had not been elected, making them vulnerable to a government challenge to the solidity of their mandate. Commitment by local authorities to a process of dialogue on limited matters gave the powerless a platform from which to demand greater concessions, using the authorities' willingness to dialogue as a lever. But in its success and its attempt to take constitutional matters into its own hands, the same process ran up against national-level politics and attempts to control.

In *Hillside,* near Fort Beaufort in the eastern Cape, where 1,000 residents had title deeds to their land from 1858 but no water, schools, electricity, or health facilities, the municipality refused any responsibility for upgrading but instead harassed the settlement in order to gain the land. The Hillside Residents' Association (HRA) refused to join another township on the other side of town in general negotiations with Fort Beaufort since it would be overshadowed by its partner, but it was also split internally between haves and have-nots. Like so many others, their only power seems to be found in their nuisance.

At *Phola Park,* a squatter settlement at Thokosa, a Johannesburg township, events over three years and still running show the breakdown of promising local negotiations and agreements triggered by state interference. Because of proximity to a white residential area, the TPA tried in 1990 to move Phola Park squatters to an area provided with sites and services 8 kilometers away. The provincial authorities imposed the duty of removing the squatters on the local authorities, thus further weakening their legitimacy. The squatters had no organization, only a new grassroots Residents' Committee that was without experience and split between Pan African Congress (PAC) and United Democratic Front (UDF) sympathizers. The Thokosa township council applied to the courts for summary removal, and the Phola Park Residents' Committee also went to court; the court sent in a mediator but the effort fell through, overtaken by problems with the police over the local hostel for migrant bachelor workers.

In September 1991, Phola Park accepted the National Peace Accord (discussed below) and joined in an agreement with the police to establish an advisory center, where the police would check in before entering the settlement. The agreement was crowned by a R25 million grant for housing development from the Independent Development Trust, a private philanthropic fund. The agreement worked for the last three months of 1991. Then the police officers were changed and, one day, the police declared that the agreement was off; it was politically not acceptable to Pretoria that the police have to give notice to enter the settlement. In reaction, a "self-defense unit" of thugs and criminals staged a "military coup," took over the advisory center, and ousted the Residents' Committee. A thousand police from the Internal Stability Units then surrounded Phola Park, set up four entry points, and checked for weapons. The self-defense unit was disarmed, but the Residents' Committee was also discredited because the people suspected collaboration with the police. By the end of the year, all alternatives for effective governance in the settlement—township council, civic, warlords, police—had been exhausted. Development (the IDT grant) was suspended. A mediator was needed that was acceptable to the parties, but the job required building structures, legitimacy, receptivity, and skills into the parties as well as mediating—a formidable task.

In many cases, including some already discussed, the parties were too inexperienced and too hostile to be able to proceed to negotiations on their own, a situation not uncommon in international as well as domestic affairs. A third party was needed to act as a catalyst and produce the necessary negotiations. Institutional mediators have arisen rapidly in the current context, brought together for the most part under the Independent Mediation

Service of South Africa (IMSSA) of Charles Nupen; in its first eight years of operations since its founding in 1983, IMSSA's activities rose to over 600 mediations and nearly 450 arbitrations. The functions of the mediator in South Africa are classic and universal: establish meeting procedures, fully air grievances, ensure parties' representativity, open communications, focus on problems and solutions, and engage parties to defend and develop a stake in their joint decisions.[5]

With or without mediation, these and other cases of community negotiations show how civics used their entrenched position and the authorities' obligation to resolve conflicts with minimum disturbance by negotiation to generate power through organization and obduracy. Inexperience, internal divisions, uncertainty, slowness, and lack of information were their greatest weaknesses. But their badgering and organizing activism, in a situation where they could no longer be ignored or eliminated, gave them power to the point where they impinged on the national actors and agenda.

Security

Security negotiations concern the politico-ethnic relations between followers of the African National Congress and the Inkatha Freedom Party. The conflict arises over positioning by the parties for a place at the ongoing national negotiations, but it exists in different forms at four different levels. While solutions exist at any level, they cannot take hold unless ratified at other levels, and yet a solution at one level is not a solution to the form that the conflict takes at another level. This poses a challenge that no local negotiations have been able to overcome, although a few have been tried at various levels.

The first level of the conflict is between the national leaders themselves, involving the positions and personalities of Nelson Mandela and Mangosuthu Buthelezi. The second is between contending political organizations, the ANC and Inkatha, where Inkatha correctly sees itself threatened by the recruiting efforts of the ANC. The ANC threatens Inkatha's position and even existence; Inkatha threatens the ANC claim of broad representativity.

But a negotiated cease-fire would also have to meet the needs at the third level of local groups, many of which have their own grievances and some of which are quite autonomous of organizational control. The conflict is tearing apart neighborhoods and workplaces, creating cycles of revenge and conditions of siege. While some of these factors serve to perpetuate and escalate the violence, others build war fatigue and a propensity to reconciliation.

But a neighborhood peace also has to cover a fourth level of conflict involving local warlords, youth gangs, and thugs who thrive on conditions of violence. In wartime, these groups are useful and their usefulness gives them a hold over the decision to pursue violence or reconciliation. At the end of 1990, neighborhood reconciliations reduced some warlords' control to a few households, but they then used their islands of control as a base from which to expand when reconciliation broke down. In 1991 and 1992, a number of local negotiations in Natal (notably Brundtville township) but also in Transvaal (notably Alexandra) stalled for lack of agreement over the place of warlords in the local delegations. Warlords are not always autonomous, and their activities play into the hands of the war faction in the top Inkatha leadership. Gangs and warlords can be brought under control only through effective and respected efforts of the police. But the effectiveness and impartiality of the police is suspect and in some cases they may even be involved with the warlords.

At the middle level of leadership, a code of conduct was negotiated by Frank Chikane and Archbishop Desmond Tutu of the South African Council of Churches, beginning in late 1988 at Cape Town and then moving to the more difficult area of Natal. By focusing on principles rather than on parties, and benefiting from the pressure of impending bloodshed, the mediators got a semblance of agreement. However, the middle-level leaders involved in setting up the code had uncertain mandates and could deliver the compliance of neither the top leadership nor the community groups or warlords. Thus, the code of conduct never became a binding pact among major organizations and fell apart when the new situation arose following Mandela's release. Its demise can be dated to the Sebokeng massacre in the Transvaal on July 22, 1990, at the hands of Inkatha.

At the second level and below, a joint working group (JWG) was set up by the United Democratic Front, Inkatha, and the Congress of South African Trade Unions (COSATU) in 1989 to handle intercommunity violence in Natal province. With the disappearance of the UDF after the unbanning of the ANC in 1990 and then the collapse of plans for the Mandela-Buthelezi meeting on March 23, 1990, it turned from a collective body into a "5+5" committee of Inkatha and COSATU, grouped in two distinct sides around Oscar Dhlomo and Alec Irwin. When these talks proved fruitless, it then evolved, under pressure from the exasperated grassroots and from international agencies (International Labour Organisation, UNICEF), into a multilevel committee based on the alliance of the ANC, COSATU, and the South African Communist Party, which sponsored two levels of activities: forums of business, church, and even Inkatha groups; and local rehabilitation committees for local truces. The JWG in any of its forms was unable to deliver the adherence of the organizations for and to which it spoke, and so it could operate neither as an effective agent nor as an effective mediator. It was only after the 1991 National Peace Accord provided a larger framework of all parties that it was able to find its place in a process.

Two lesser agreements were negotiated within the JWG context but at the local level. The Mpumalanga township agreement was a truce brokered by local businesspeople and shop stewards to stop the most violent war in Natal. The conflict left the community exasperated, producing grassroots support for an agreement; business was threatening to leave the area as a result of the conflict. The truce ended the war in the factories, and the police and SADF enforced it in the township, under the surveillance of the press and diplomatic observers. However, because it was not locked into place with an agreement on the first and fourth levels, committing the national leaders and restraining local warlords, the truce broke down in early 1990. A second attempt to restore peace to Mpumalanga township was undertaken by local ANC and Inkatha leaders two years later, and it has shown success in deescalating violence and initiating reconstruction. The participants had simply had enough.

In Imbasi, a local rehabilitation committee arranged a similar truce among warring and ill-controlled factions in the suburb of Pietermaritzburg in 1991 by depoliticizing the conflict and harnessing local resentment at continuing insecurity. By trading insecurity for insecurity, and then truce for truce, and by using local business resources as an incentive, the mediators were able to restore law and order. But the local mediator was threatened with assassination, probably by Inkatha elements, and the local truce collapsed. Business became skittish about risking its resources, the participating organizations did not cooperate, and resentment and suspicion rose against the peace initiative. The experience shows that a mediated truce is only the first step and requires firm follow-through to stay in effect.

Security negotiations are possible at various local levels; paradoxically, by capitalizing on local community exasperation at rampant insecurity, they eliminate the very incentive that makes their outcome attractive. After a time, the lack of buttressing agreements at higher and lower levels destroys the progress achieved, and insecurity returns. The resulting situation is actually worse than before, since negotiation has been discredited in the process. Nonetheless, local attempts at security agreements were a major contributor to the negotiation of the National Peace Accord[6] among the South African government, the ANC, the IFP, the labor unions, of many other parties, and religious and civic organizations. Signed on September 15, 1991, this agreement was the culmination of local efforts that has then helped to hold the local agreements in place.

Education

Negotiations on education are in a paradoxical situation. It was an educational crisis that triggered the school boycott of 1985 and the state of emergency for the rest of the decade. The National Education Crisis Committee (NECC) and various school committees of parents, teachers, and students negotiated with school administrators in 1986. Yet these negotiations accomplished little; the authorities were less vulnerable to student disorders than they were to squatter disorders, and to school boycotts than to rent boycotts. As a result, education negotiations in the 1990s have thus far addressed traditional student concerns more than the much needed constitutional changes. An SPD leader has identified education as "the next issue."

Two cases are not exhaustive but rather indicate the limits of the current education negotiations. In late 1989, at the University of the Witwatersrand, the two student organizations—the black South African National Student Congress (SANSCO) and the white Student Representatives Council—joined in a national campaign to protest the termination of students who fail to meet university standards. The vice-chancellor's reply was not deemed satisfactory, and in February 1990, a three-day strike alleging inadequacies in remedial counseling, funding, and housing, and calling for a moratorium on terminations attracted a small number of students. The university then offered to constitute a special review panel of two NECC members and two university senators to examine the twenty-nine cases of termination; twenty-two were rejected, three were reaccepted (one declined), and four were to be reconsidered in 1991.

The response was a precise answer to a general grievance; a university spokesperson noted that "the university had to invent a solution to get the students off the hook." The university also found a new housing facility, although not nearby. Both the termination and the housing issues were complicated by racial implications, and the demands posed by student organizations (backed by COSATU) pointed to a "1960s-type student and workers governance," as the university put it. The negotiations were more typical of a normal campus than of a national regime change; they enlisted little student support and raised no power issues, despite some intentions to the contrary. But they did not address an underlying issue deep in the matter of regime change: the clash between standards and the demands made on students unprepared by the apartheid system of education to meet those standards for admission or completion of their studies. A similar round of demonstrations and negotiations also took place at the University of Cape Town, where students were invited to be part of the review committee on readmittances but refused.

The other case concerns the integration and privatization of Bernardo Park girls' high school in order to keep it open. Population aging and depletion in the neighborhood dropped the all-white student body to less than a third of its former size of seven hundred, while the rise of the African population in the area brought individual requests to enter the school. The situation, not the parties, made for conflict, and there was no debate over the curriculum. There was no organized pressure on the black side, and a small group of leaders acting as a "board of trustees" of the school acted for the black parents; the white parents' organization, Save Our School, did not oppose the integration and was more interested in keeping the school open. The conflict arose over the status of the school and hence its financial position. The state could not run an integrated school, but the school was not in a position to operate as a private institution. By the White Only Schools Act, as amended on June 29, 1990, registered schools had to be more than 50 percent white; Bernardo Park's student body of two hundred sixty was only 40 percent white, and Africans constituted the largest group. Negotiations were conducted among the parents and with the school administration and were successful in keeping the school open and integrated in 1990; negotiations were also conducted with the government and were unsuccessful in resolving the legal anomaly, despite agreement all around that the problem had to be met and was growing. The legal contradictions are typical of the apartheid system's operations in many sectors of life, and of the need for constitutional resolutions.

Educational negotiations are still not at the stage of direct action and confrontation of the constitutional issues that are typical of community negotiations. As a result, ad hoc solutions are worked out among consenting adults, while the authorities proclaim their impotence in the absence of a constitutional resolution. The question of power does not arise in the cooperative negotiations and is inadequate in the confrontation with the authorities.

Labor

Labor negotiations rose significantly first in the 1980s and then again after 1990 as the political system opened up and prospects for codetermination (if not self-determination) suddenly improved. Labor, like local communities, saw not only the possibility for taking pieces of its destiny in its hands, but also the immediate need to assert its role among the players as major negotiations became imminent. But labor was also under pressure from its members, whose expectations rose enormously after the events of February 1990 and then the constitutional negotiations in the Conference for a Democratic South Africa (Codesa) in 1992, impelling union leadership into greater challenges but also bearing an implicit threat of more radical leadership if the incumbent moderates were not able to produce results.

Labor's source of power stemmed from that threat, but more fundamentally from one of the most important early reform measures that emerged from the Botha regime: the legalization of black trade unions in 1979. This established the framework for institutionalized negotiations on social as well as economic issues between black workers and white employers, ultimately with deep constitutional implications. Legislation circumvents labor power, however, by regulating strike conditions, prohibiting strike funds, and allowing blacklisting, repatriation, and vagrancy arrests. Thus, despite the legalization of unions, labor negotiations are as asymmetrical as communal negotiations.

At one extreme, P. G. Byson in Johannesburg is a holding company of ten firms and a member of the Cooperative Business Movement founded in 1987 and devoted to good labor relations. The company seeks to avoid strikes and confrontational negotiations, but rather attempts to expand information and develop cooperation. Wage negotiations began in May 1990 with management first issuing information to the union about the financial condition of the company. It then offered an increase of 17 percent, one point below the top possibility in the budget. The offer was rejected for Boulwarism,[7] and the workers held solidarity demonstrations—including *toitoiing* (traditional conflict dancing)—outside the factory at lunchtime. Negotiations were resumed in July, in a five-hour session, and ended with an agreement at 17.6 percent. Shop stewards are the active agents of negotiations, with the support of the workers, and are not hesitant about demanding information or rejecting offers; COSATU warned the stewards about "bosses undercutting labor militancy" during the negotiations and the stewards rejected that criticism too. The company reports that its settlements without strikes tend to be about equal to or slightly below fellow companies' settlements after strikes.

Retail workers' negotiations in mid-1990 between the South African Commercial, Catering and Allied Workers' Union (SACCAWU) and Checkers, OK, Pick & Pay, Trador, and Metro Cash & Carry, among others, were an important hallmark of the new labor confrontations of the new decade. Each company has its personality and way of dealing with its employees, and the fate of each set of negotiations was determined in part by its place in the series. The negotiations have been crucial because they mark an interplay of sociopolitical and economic goals. Wage demands skirt the edge of the 8 percent profit margin, while at the same time defining and reshaping the future shapes of labor-management relations in the new order.

On its side, SACCAWU had just emerged in new form and under new leadership in its November 1989 congress. It was dominated by a new outspoken leader, Jeremy Daphne, who had trumpeted the targets of R160 raises and R800 minimum wages as the union goals. But the union was still torn by horizontal and vertical divisions not only between ANC and Pan African Congress (PAC) factions but also between national leadership, politically versus professionally motivated shop stewards, and rank and file workers. The union's goal was to return to an industrial council arrangement, where SACCAWU would negotiate with all retailers together. Such an arrangement had existed before the legalization of black trade unions, with white labor negotiating with white management and then applying the agreement to blacks. Now management continued to reject the industrial council model as inimical to business competition and as conducive to the practice of "second-tier bargaining," where unions returned three months after a generalized agreement to leapfrog improvements on a company basis. SACCAWU also launched a broad labor relations program as part of its new existence, calling for a cradle-to-the-grave plan in June 1990, endorsing union solidarity at its conference in mid-July, and at the end of the month holding a one-day work stoppage for the right to strike and picket without punishment. Its wage negotiations took place in the midst of these campaigns, which strengthened both its image and its position.

In the early settlements in July 1990, Pick & Pay, after a twenty-four-hour strike, reached the first agreement on an across-the-board increase of R160, which then became the benchmark figure. Trador followed with the same figure; it had sought close relations with its employees, fed them during the strike, and agreed to a moratorium on layoffs for eighteen months, even though it had planned to close two stores. The two key companies were OK and Checkers, two very different companies. Checkers sought good, interactive labor relations but without the extremes of Trador or Byson; it had never had a strike but was blamed by its workers for bad management. OK was an old-time hardline firm, operating on a low profit margin, high dividends, and little plowback into company development; it had a major, costly retail strike in 1987.

OK and SACCAWU began negotiations in February 1990 with the aim of reaching agreement in early April. After initial discussions on the scope of the talks and the topics at issue, with the mediator (IMSSA) suggesting exchanges of information and fairness guidelines as a beginning, the talks bogged down. By April the parties were at R115 and R160, respectively; the union asked for a two-week reflection period but instead began preparing for a strike. After two weeks of strike, the company offered stepped longevity increases (from R115 to R145 for every five years' service), which won agreement from the two-thirds of the employees who were nonunion, from a third of SACCAWU members, and from another small union. OK then used police and dismissals against the remaining strikers. The union responded with R150; the company countered with three longevity grades (R125–R145) instead of five. In mid-July, the mediator brought the union to face the real costs of a continuing strike, and then brought the company to make an offer the union would be able to accept. OK then added a new holiday, raised the minimum from R720 to R730, increased each longevity grade by R5, but also included plans for retrenchment and punishment of strikers. An additional R5 was added to the grades and the deal was struck, at less than the benchmark and other than across the board. The pattern of movement was a typical labor-management concession/convergence dynamic, but the key, wielded by the mediator, was the vulnerability of each party to the power of the other.

Checkers also negotiated during the period and by mid-July had moved from an R60 to an R5 gap. On July 16, Checkers issued a take-it-or-leave-it offer until ten o'clock the next morning, but SACCAWU locked the negotiators in the room until they agreed to continue negotiations. Once agreement had been reached at R145 and R770, the union added on more details (some rather large loopholes), which management sharpened to acceptable points; and then it changed negotiators, bringing in Daphne himself at the last minute. Agreement was held up by the union leader for twenty-four hours, in order—as it turned out—to sign at the same time as OK.

The 1990 retail negotiations were tests of strength between an aggressive union locked in its policies by political conditions and a management on the defensive politically despite narrow economic margins. Beyond raises and minima, it got into other economic details such as bonuses and discounts and political conditions such as new holidays and retaliation. Still outstanding were joint determination issues such as pension fund management, a growing—and constitutional—issue for unions in mining and manufacture. Retail workers were able to use the strike as their major weapon, but were unable to go further into consumer boycotts for the issues at hand; they were able, however, to play on the competitive nature of the industry and its narrow profit margins and turn them to its advantage.

The final case is a fascinating foreshadow of the new order, constitutional in depth but local only in the sense that it was a sectorial, economic negotiation rather than a broad political one. It concerned amendments to the Labor Relations Act (LRA). The previous round of intense labor union activity in 1987 led to a series of government amendments to the act under pressure from some business sectors to recover initiative from the unions and strengthen business in its dealing with the unions. At the same time, some management groups—notably the Federation of Chambers of Industry, some members of the National Manpower Commission, and eventually the South African Employers' Co-ordinating Council on Labour Affairs (SACCOLA)—felt that it was time to initiate direct and comprehensive business-labor talks to deal with the current pressures of both job actions and sanctions. Talks began in March 1988 between SACCOLA and the two labor federations, the National Council of Trade Unions (NACTU), close to the PAC, and COSATU, close to the ANC. Half a dozen members of each of the three organizations met to discuss labor issues for the first time. However, COSATU at its congress in mid-May decided on a three-day strike the following month in protest against the impending amendments, seen as a result of business pressures. Instead, a delegation of SACCOLA representatives and labor lawyers met with the manpower minister throughout July and August and obtained agreement on withholding six particularly offensive articles (on strike limitations, use of courts, time limits, and unfair labor practices, among others). However, despite an agreement announced on August 11 not to promulgate the law, it was passed by parliament and issued on September 1, the minister apparently having been miffed at the unwillingness of labor—rather than its lawyers—to meet and discuss with government. The unions, in a labor-based sense of the way organizations work, then asked SACCOLA to in-struct its members not to apply the offending articles, but SACCOLA had no such power. The first round of negotiations broke down over erroneous expectations on all sides.

In mid-1989, the talks revived under union initiative and by May 1990 had reached an agreement on changes in the law. In mid-March, the man-power minister had promised the three parties that any agreement among them would be taken into account by government; most of the agreement (except for provisions on labor courts and unfair labor practices) was later endorsed by the National Manpower Commission. Throughout the negoti-ations, the parties again had problems establishing their mandate and en-gaging their constituencies. But a further problem arose when the labor unions wanted to take advantage of the new negotiations and extend the provisions of the law to civil servants, currently not covered by the LRA, thus bringing government into a tripartite consultation with labor and man-agement. National Manpower Commissioner Fourie demurred, stating that government could not be both a party and an adjudicator, a reflection of government's position on the upcoming constitutional negotiations. In the end, the proposal was submitted to parliament in September and enacted in early 1991.

The LRA labor-management consultations were an extraordinary ex-ercise of unofficial—even if not local—negotiations. Far outreaching the impact of normal labor-management bargaining, even when these extend beyond wage and productivity issues, the negotiations were clearly consti-tutional in nature. Marked by stay-aways, walkouts, and external diver-sions to add pressure, they tested the ability of the basic elements of soci-ety to come to terms on a social contract. The potential power of both sides was tremendous, too large to be wielded effectively in the negotia-tions but large enough to contemplate in the danger of breakdown. In the process, the parties learned to talk to each other and to communicate and negotiate in good faith, beginning with overcoming the objections of NACTU to participating at all.

Implications of Local Negotiations

There are a number of apparent patterns to these negotiations. First, they follow a standard pattern of "petition politics," in which social groups pose demands to the authorities and then put on the pressure, with threats of greater disruption, until a response is forthcoming. This is petition politics rather than "normal politics" because the authorities are in no way responsible to the social groups, since the majority has no vote in South Africa. It is different from standard negotiations in that the parties are not equal and do not engage in exchange of benefits; the only—but basic—trade-off underlying the negotiations is the exchange of responses for the release of pressure. The pressure can take the form of a demonstration, picket, strike, boycott, etc.—some form of denial and disruption that constitutes the only means by which the petitioning group can redress the power imbalance.

Yet how does the social group get the power to pose a credible petition, or, put otherwise, what elements of power do authorities have and petitioners not have that make for asymmetry and that need to be overcome to escape from it? Four such elements can be identified: mandate, standing, organization, and information. Workers, squatters, community residents, parents, and students find that they need to develop a means of representation and accountability toward their client groups, achieve recognition from governmental authorities, provide structure and continuity for their support, and inform themselves on the technical details of their subject matter so that they will not be dependent on the expertise of government. If they do not do these four things, they will be too vulnerable to petition effectively. The need for empowerment or consolidation is a precondition to local negotiations, and it often requires that negotiation be suspended until it is achieved; yet to the extent that consolidation is achieved and a response is obtained, the petitioner emerges from each encounter empowered by the very process of negotiation.

Empowerment emerges from the content as well as the conduct of local negotiations. While all local negotiations begin as petitioning, new issues inevitably are added to move the local group from a petitioner to a codeterminator. The increase in power that accompanies a shift from a problem poser to a problem maker raises the party in status to a (joint) problem solver. When a union becomes a comanager of a pension fund, when workers or their representatives become directors of a trust fund, when townships achieve a single tax base with cities, when parent-teacher associations decide on new curricula or enrollments, the parties have moved from petition-and-response to joint decisionmaking or full negotiations, or from substantive to constitutional issues.

Beyond petition and empowerment is the establishment of relationships between local and national negotiations. There is no doubt that local negotiations parallel and exist independently of national negotiations. Both occurred during the second half of the 1980s, on totally distinct issues; both have increased dramatically in importance in the 1990s, with convergence on both process and issues, but still occur autonomously. Yet as the two levels become more open and expand their subject matter, they become necessarily linked, in a number of different ways. First, empowerment—growth in standing, mandate, organization, and information—at the bottom empowers the top. This is particularly true when there is an organizational link between the two levels, but it is true too even when unions, civics, parent-teacher associations, and other groups remain organically distinct from the ANC.

Second, local negotiations reinforce the national process. To the extent to which local petitioners demand responses from and then make joint decisions with local authorities, they work to make participatory rather than authoritarian decisionmaking the dominant mode of governance. Unless rooted at the bottom, a change at the top in the mode of governance has no chance of survival.

Third, local negotiations democratize national negotiations. While they do not remove the possibility of the elite deal that figures so prominently in the democratic transition literature from Latin America and Southern Europe,[8] they provide an active basis for any deal that evolves, and a deal of their own that keeps the upper level from drifting too far away from the base. In addition to wrestling with local issues, usually of national import, local negotiations also have their procedural agenda as a struggle between organizations of top-down versus bottom-up politics, providing a vigor to the lower dimension that cannot be ignored. In the process, attempts to control the civics or the locals can also disrupt negotiations, destroy initiative, and even open the way for a breakdown of local patterns of politics and authority. That is to be expected, for the whole South African evolution in the 1990s involves a sudden breakdown and reconstruction of politics and authority. Local processes are important ingredients in that accelerated evolution.

Fourth, talking at all levels tends to replace fighting. At the beginning, pressure, including supportive use of violence, is part of the negotiating process. Here as elsewhere, it would be a mistake to think that the two are mutually exclusive alternatives rather than mutually supportive means. But as negotiation becomes the accepted mode, violent pressure becomes secondary or no longer necessary. In order to become the accepted mode, negotiation must be tried, tested, and eventually institutionalized, at all levels.

Fifth, some issues begin to be decided at the local level, locking the nationals into jointly determined results. The common tax base issue and the all-municipal councils are an important example, so important that local moves have sometimes been delayed for national moves to catch up with them. When the Soweto People's Delegation and the Transvaal Provincial Authority agreed to study jointly implementation of the SPD's five demands, the first steps were being taken in local negotiations to legislate a common tax base and a common urban area as a national policy;

other municipal areas added to the pressure in their own terms. One of the most striking cases of evolving local negotiations is the COSATU-NACTU-SACCOLA agreement on amendments to the Labor Relations Act, which grew out of individual labor-management negotiations; of it, a leading COSATU spokesperson said, "We want to get a Labor Relations Act that a new ANC government will have to live with." Unless local negotiations confront aberrations in labor relations practices, or expose the unworkable city-and-township system, or bring to light the contradiction between standards and education, or show up the police role as well as the national rivalries in local insecurity, the pressure for change in national legislation would not be present, or at least not as compelling. It is noteworthy that in the process, "pressure" changes meaning from the local to the national level as the process of negotiation expands.

Finally, as some aspects of the previous two examples also suggest, local negotiations have often taken an issue as far as they could and then, running into national legislation, have kicked it upstairs for a legal solution. Local negotiations are limited in what they can do; they can run into legal barriers to the implementation of their decisions and face deadlock. This deadlock is used tactically by the central government on occasion as a means of discrediting the civics and the local participation process. Civics and other local actors then refuse to sign because they cannot deliver what they have decided. Deadlock creates frustrations, turning labor, education, and housing disputes into security conflicts. Thus, local negotiations alone are not the solution; the new system cannot be remade from the bottom up. But in the last analysis, even frustration of the process in some local instances creates pressure for a national solution, with meaningful local participation.

Notes

I am grateful for the helpful comments of Sipho Shezi, Mark Anstey, and Thomas Karis on this chapter. I am also grateful to the MacArthur Foundation and the Carnegie Corporation for support for the research on this topic, to the Human Sciences Research Council and Louise Nieuwenmeijer for a timely conference, and to the United States–South Africa Leadership Exchange Program (USSALEP) for its assistance.

1. There is an impressive literature in South Africa on this subject. See Mark Swilling, *Local Level Negotiations: Case Studies and Implications* (Johannesburg: Urban Foundation, 1987) and *Beyond Ungovernability: Township Politics and Local Level Negotiations* (Johannesburg: University of Witwatersrand Centre for Policy Studies, 1989); three books by Paul Hendler, all published by the South African Institute of Race Relations in Johannesburg: *Urban Policy and Housing* (1988), *Paths to Power* (1989), and *Politics on the Home Front* (1989); and Mark Anstey, *Negotiating Conflict* (Cape Town: Juta, 1991), *Practical Peacemaking*

(Cape Town: Juta, 1993), and "Mediation in the South African Transition," *Genève-Afrique* 30, no. 2 (1992), pp. 141–163.

2. See Jeffrey Rubin and Bert Brown, *The Social Psychology of Bargaining and Negotiation* (New York: Academic Press, 1975); I. William Zartman, *The Politics of Trade Negotiation Between Africa and the EC* (Princeton: Princeton University Press, 1971); I. William Zartman, ed., *Positive Sum: Improving North–South Negotiations* (New Brunswick, N.J.: Transaction, 1987); Mark Habeeb, *Power and Tactics in International Negotiation* (Baltimore: Johns Hopkins University Press, 1991); Jeffrey Rubin and I. William Zartman, eds., *Power and Asymmetry in International Negotiation* (Laxenburg, Austria: International Institute of Applied Systems Analysis, forthcoming).

3. This is a study in oral history, for which there are few written records other than internal documentation of the parties. It is based on over seventy interviews conducted in the Johannesburg, Pretoria, Pietermaritzburg, Natal, Port Elizabeth, Grahamstown, and Cape Town areas in 1990 and 1992.

4. See Khehla Shubane, *The Soweto Rent Boycott,* B.A. Honors diss., Political Science Department, University of the Witwatersrand, 1987; Mark Swilling, *The Soweto Rent Boycott* (Johannesburg: PlanAct, 1989).

5. On the universalities of mediation, see Kenneth Kressel and Dean Pruitt, eds., *Mediation Research* (San Francisco: Jossey-Bass, 1991); in South Africa, see Anstey, *Practical Peacekeeping.*

6. The text has been published by the UN Centre Against Apartheid, Notes and Documents 23/91, November 1991. For another successful peace initiative in local mediation, see Khotso Kekana and Albert Nolan, "Making Peace from Below," *Challenge* 14:2–4 (April 1993).

7. Named after the tactic of a GE official who in the 1950s made fair offers to labor as ultimatums, and was rejected for taking the participation out of collective bargaining. See David Lax and James Sebenius, *The Manager as Negotiator* (New York: Free Press, 1986), p. 235.

8. Guillermo O'Donnell and Philippe Schmitter, *Transitions from Authoritarian Rule* (Baltimore: Johns Hopkins University Press, 1986).

6

Dislodging the Boulder: South African Women and Democratic Transformation

Amy Biehl

On August 9, 1956, 20,000 South African women marched in Pretoria and threw their passbooks at the feet of Prime Minister Strydom. As they marched, they sang:

> *Wathint'abafazi. Wathint'mbokotho. Uza kufa.*
> [You have struck the women. You have dislodged a boulder. You will be crushed.][1]

Nearly forty years later, on April 1, 1993, women marched to the World Trade Centre in Johannesburg to demand a central role in the national negotiations forum. The ANC Women's League secretary-general, Baleka Kgositsile, reminded the negotiators that women were more than 53 percent of South Africa's voting population. "No stone will be left unturned in the battle for the rights of the majority," she said.[2] For South African women, the World Trade Centre protest marked a change in approach from mass resistance to strategic inclusion. Nevertheless, as in 1956, the message was clear—the boulder had been dislodged. After decades of resisting apartheid, women would not be excluded from negotiations for a democratic government.

Between 1956 and 1993, the South African political climate changed dramatically from resistance and armed struggle to negotiations and political compromise. For nearly three decades, women's liberation had been considered secondary to national liberation. However, according to Cherryl Walker, "Today, as a victory over apartheid moves into the realm of feasible rather than utopian politics, so, finally, more serious attention is being given to what the most appropriate ordering of the relationship between women's and national liberation should be."[3]

The failures of women to benefit equally from national liberation elsewhere in Africa helped spark a debate over women's rights both within the Mass Democratic Movement (MDM) inside the country and within the ANC in exile. Organizations such as the United Women's Organisation (UWO) of the Western Cape played an important role in the formation of the United Democratic Front during the 1980s. Several conferences and forums were convened outside of the country prior to and during 1990 to discuss gender within the liberation movement, including an ANC seminar in Lusaka in 1989 and the joint ANC-MDM Malibongwe conference in Amsterdam in 1990.

The unbanning of the ANC in 1990 and the beginning of the negotiations process changed the rules of liberation politics. Mass resistance, which had relied heavily on the support of women, changed to "closed-door" negotiations dominated by male leaders. For women, this change has not been easy to make for many reasons: vast differences among women themselves, a culture of patriarchy that crosses all political lines, and a culture of violence engulfing the country. Nevertheless, women from all social and political backgrounds agree that the negotiation process presents a historic opportunity for women to help shape the form and content of a new democracy and to develop long-term strategies for transformation.

This chapter will (1) describe the current legal, demographic, and socioeconomic position of South African women; (2) outline strategic issues for women in the transition as well as in the long-term process of democratic transformation; (3) discuss the impediments and challenges to women's success in this process; and (4) discuss possible strategies for change.

Profile of South African Women

There are 10,188,521 adult women in South Africa, excluding the TBVC areas (the "independent" homelands of Transkei, Bophuthatswana, Venda, and Ciskei).[4] Of these women, 70 percent are African, 16 percent are white, 11 percent are coloured, and 3 percent are Indian. Of all women (excluding the TBVC areas), 38 percent are located in the "self-governing territories" of Gazankulu, KaNgwane, KwaNdebele, KwaZulu, Lebowa, and QwaQwa. The exclusion of the TBVC areas from census figures makes it difficult to obtain accurate numbers, but the Development Bank of South Africa estimates that one in eight South African adults, and one-fifth of the African population, live in the TBVC areas. Many more women than men live in the TBVC areas as a result of male migrant labor caused by the apartheid system. An estimated 2,105,000 adult women live in the TBVC areas.

Thirty-four percent of South African women (excluding the TBVC areas) and 41 percent in the TBVC areas are under the age of fifteen. Thus, while South African women currently make up approximately 54 percent of the potential voters in the first nonracial elections, by the turn of the century, women will be more than 60 percent of the voting population.[5] Over half of the African women (excluding the TBVC areas) live in rural areas, while 82 percent coloured, 92 percent white, and 96 percent Indian women live in urban areas.

In terms of education, women are disadvantaged in comparison to men only at the highest levels of education. However, among women of different race groups, significant disparities in literacy occur. For example, 99 percent of white women thirteen years or older have at least a Standard IV (sixth-seventh grade) education, while 1989 figures for the Transkei show that one-third of all women over the age of five have no education at all and only 15 percent have some secondary education. An estimated 3 million women in South Africa are functionally illiterate. More than three-fifths of the 542,713 women with postmatric (university level) degrees (excluding the TBVC areas) are white.

Approximately 41 percent of adult women in South Africa are married. When analyzed by race, 59 percent of white women are married, 57 percent of Indian women, 40 percent of coloured women, and 35 percent of African women. However, statistics on the marital status of South African women are extremely difficult to calculate because several different marital regimes can apply.

Marriages can be recognized as civil marriages or as customary marriages. For civil marriages, coloured, Indian, or white women married before November 1, 1984, and African women married before December 2, 1988, are considered legal minors. These women are required to have permission from their husbands to open bank accounts, sign property leases, and conduct similar legal transactions. Women married after the change in marital laws are considered adults before the law. However, in all South African marriages the husband has "marital power." Marital power allows the husband to decide "where and how the family will live," including issues such as the custody of children in the event of divorce or separation.[6] A marriage can take place under ante-nuptial contract, specifying the legal power of each spouse. Civil marriages can occur in community of property, out of community of property, or by ante-nuptial contract with accrual, which determines the division of property in the event of the death of one spouse or divorce. Under South African law, a woman cannot charge her husband with rape.

Two types of customary marriages are recognized in South Africa: African customary marriages and Muslim marriages.[7] Outside of KwaZulu or Natal, women married by African customary marriage are considered minors. In KwaZulu and Natal, the husband has marital power similar to that under current civil law. Under Muslim marriages, women are considered full adults. Under customary law, a married woman has some rights, including a right to her husband's pension in the event of his death and the right to claim maintenance for children, if she can prove the marriage with a certificate or by demonstrating that *lobola* (a form of bride price) was paid. However, most women in customary marriages are unaware of these limited rights. Under the inheritance provisions of customary law, most of the husband's property goes to the men in his family. Prior to December 1988, if a husband took a second wife by civil marriage, the date of the civil marriage officially terminated the customary marriage.

Excluding the TBVC areas, women represent 36 percent of the workforce, but only 11.5 percent are found in mid-level management positions and 6.7 percent at top-level positions.[8] In the TBVC areas, an average of 9 percent are classified as economically active. However, statisticians do not regard women engaged in subsistence agriculture as economically active. Of economically active women, white and Indian women are found mostly in clerical, sales, professional, and technical positions. Coloured women work in service, clerical, sales, and supervisory positions, but in fewer technical and professional positions. African women are found mostly in service positions and in the agricultural sector. Over 95 percent of the few African women in professional occupations are teachers or nurses.[9] Most African women and most coloured women in the service profession are domestic workers.

More than 70 percent of South African women (excluding the TBVC areas) report having no income at all. This includes more than 7 percent of those women classified as economically active. Eleven percent earn between R1 (33 cents) and R4,999 ($1,675) annually. Fewer than 2 percent of each of the three "black" groups earn more than R10,000 ($3,330) a year, compared to 9 percent of white women. Despite their extremely low incomes, large numbers of African women are considered de facto heads of household. It is estimated that nearly half of working women (excluding the TBVC areas) are single mothers and heads of household; the percentage rises as high as 59 percent in rural homelands. The only child care system that exists for working mothers (who can't afford domestic labor!) has been a result of informal networking among women in townships. Many working mothers send their children to rural areas to be cared for by relatives.

Violence against women has reached crisis proportions in South Africa. Rape Crisis Cape Town estimates that a rape occurs every 83 seconds. One in every six adult women is regularly assaulted by her partner.[10] In 1992, 218 women were killed and 251 women injured in "politically motivated" violence.[11] Women made up twenty-five out of forty victims identified at the 1992 Boipatong massacre, which led to the breakdown of negotiations later that year.

There are an estimated 300,000 illegal abortions performed each year.[12] Thousands more women die each year as a result of backstreet abortions. As a result of existing marital laws, many hospitals require a husband's signature before they will permit surgical procedures for married women.[13] Overall, women's specific health issues have been inadequately addressed by the state, and community health projects designed to assist women have been underfunded and overstretched.

The above statistics demonstrate that South African women are incredibly diverse in terms of race, income, literacy, and other socioeconomic indicators. The majority are economically dependent on their husbands or relatives, and those women who work do so in the least desirable positions, earning the lowest incomes. Black women are particularly at risk in terms of economic survival. Women of all races and income levels are threatened by physical violence. Women have little control over their own reproductive freedom and marital relationships. These factors will shape the strategies women will pursue in the transition and their goals in the broader process of transformation.

The current position of women in South African society requires that consideration be given to (1) strategic issues of political participation and the establishment of formal gender equality during the transition process, and (2) practical issues, such as access to employment and physical safety from violence during the broader process of democratic transformation. A combination of these two approaches is necessary if women are to participate fully in, and benefit equally from, the transition to a democratic society. Whether strategic or practical, mechanisms for women's empowerment must recognize the intersection of race, class, and gender as it affects South African women. Strategies to remedy discrimination must not only reflect these different experiences, but they must emphasize the empowerment of the rural woman living in the former homelands if they are to reach to the "average" South African woman.

Women's Political Activity: 1990–1993

The unbanning of the ANC and other liberation organizations in February 1990 changed the nature of women's activities. Women who had been in exile returned to join women who had remained in the country. Forums such as the 1990 Malibongwe conference helped smooth this transition process for women within the liberation movement and helped them define an agenda consistent with the new situation. Women adapted to these circumstances much more quickly than their male counterparts, recognizing an important political opportunity to make their views heard. Several women's groupings, such as the Women's Alliance of the Western Cape and the Women's Alliance of Southern Natal, organized efforts to forge links among women, collect demands for a women's charter, and campaign for women's full equality.

Alarmed by the lack of women negotiators at Codesa, women proposed that a mechanism be established to ensure their participation. The Gender Advisory Committee (GAC) was subsequently established in 1992 as a subcommittee of the Codesa management committee. Coordinated primarily by ANC and DP women, the GAC was in fact one of the few committees at Codesa II that reached consensus, and its members chose to retain the GAC's *locus standi* after Codesa was disbanded.[14]

Despite important progress made by different women's organizations and the GAC, many women expressed the need for a national women's organization that could enjoin women from the broadest possible spectrum of political organizations to articulate women's demands. As a result, in April 1992, the Women's National Coalition (WNC) was launched. Approximately seventy organizations were represented and more than sixty organizations have since confirmed membership. Regional coalitions currently exist in eight regions of the country and more are expected to be initiated. WNC members include the ANC Women's League, the Democratic Party Women's Forum, the African Women's Organization of the PAC, the National Women's Chapter of the National Party, the Women's Wing of the Disabled People of South Africa, religious organizations, occupational and professional groups, advocacy groups, service groups, and special interests groups, among others.

According to WNC Director Pregs Govender,

> The WNC was formed to unify women in formulating and adopting a Charter entrenching equality for women in South Africa's new constitution. It plans to do this through a process aimed at empowering women. The Women's Charter Campaign consists of an organized and systematic exercise to do the above. . . . A multi-method programme of participatory research will be an integral part of the whole project.[15]

Although originally given a period of one year, the steering committee of the coalition has extended this period to accommodate the changing transition process.

Despite its successful launch, difficulties of working across vast political differences, inadequate infrastructure, and problems with facilitating regional activities have plagued the WNC. Many organizations, such as Black Sash and COSATU, have expressed concern that their own constitutions will not permit them to affiliate with the WNC. The limited mandate of the WNC has angered some members who would like to address national issues such as violence through the coalition. At the WNC Council meeting in February 1993, the national convener, Frene Ginwala of the ANC, acknowledged that the expectations of the coalition had "been based on hope rather than reality," but the WNC revised its program, strengthened its infrastructure, and agreed to continue to "work across differences to achieve the WNC goals."[16]

Perhaps the most serious test of the WNC to date was the debate sparked by a National Party proposal for three draft bills on women's equality: the Abolition of Discrimination Against Women bill; the Prevention of Domestic Violence bill; and the Promotion of Equal Opportunities bill. The bills had been drawn up by a junior advocate in the Ministry of Justice with minimal consultation with women's organizations. Most important, the WNC was excluded from the consultation process.

Women in the NP considered the bills an important victory. Outside the NP, however, most women were extremely dissatisfied with both the content of the bills and the unilateral method by which they were introduced. Democratic Party MPs Dene Smuts and Carole Charlewood argued in parliament that the draft legislation was an inadequate attempt to duplicate the participatory process initiated by the WNC, and that it had occurred at taxpayers' expense. At a conference convened by the Ministry of Justice to explain and promote the bills, Professor June Sinclair received a standing ovation for her scathing review of the proposed legislation, in which she argued that the omissions in the legislation, including the disregard for black women and the avoidance of abortion law reform and tax equity for women, represented more glaring policy statements than the legislation itself.[17]

The government proposed that women and women's organizations be given approximately one month to submit comments on the legislation. While some organizations provided significant commentary on the legislation, many groups, including the WNC, rejected the process and refused to comment. These activities became a divisive force within the WNC. Women of the liberation movement believed that NP women had used the coalition as a vehicle to preempt the goals of the WNC, while NP women believed that the "ANC Charter Campaign" (WNC) was preventing women from capitalizing on an important window of opportunity presented by the legislation. Nevertheless, the WNC agreed to differ on the issue of the government's legislation and proceeded with the Charter Campaign.

The current activities of women and women's organizations can be classified as focusing on one of three phases in the transition/transformation process: (1) from the current negotiations to elections for a constituent assembly/new parliament, (2) the constitution-writing process, and (3) the Government of National Unity and Reconstruction and beyond. Some issues intersect and overlap with these different phases. However, a discussion of key strategic elements with regard to each phase follows.

Current Negotiations

The breakdown of Codesa destroyed important progress made by women to increase their representation through the GAC at Codesa II. The bilateral talks that took place following the breakdown of Codesa II virtually excluded women. Subsequently, the reconstituted World Trade Centre talks began as nearly an all-male forum. The bilateral meetings also brought a greater number of conservative and fundamentalist organizations and traditional leaders into negotiations. In addition to the main negotiating forum, women have also been excluded from the peace forums, economic forums, local government forums, and other parallel negotiating structures.

In March 1993, the former GAC put forth a proposal to revive the GAC at the World Trade Centre talks. Women in the Democratic Party proposed that the GAC be given subcommittee status and an "input function" to the negotiations. The ANC Women's League felt that "subcommittee status" for the GAC was insufficient and lobbied their organization for an ANC proposal to amend a pending proposal that the Negotiating Council consist of one negotiator and one adviser from each party. The ANC proposed that this be changed to one negotiator and two advisers, and that at least one of the three representatives per party be a woman.

This proposal, put forth by ANC negotiator Cyril Ramaphosa, was booed and jeered by the negotiators present. Only three parties of the twenty-six—the ANC, the SACP, and the Natal Indian Congress/Transvaal Indian Congress—supported the proposal. As a result, the ANC women caucused with other women's organizations to protest their exclusion. Following these meetings, the IFP Women's Brigade came forth with an even stronger approach. It rejected the reconstitution of the GAC and proposed that all party delegations and committees be expanded to add one woman delegate, who would have full negotiating rights. Women's efforts resulted in the final wording, agreed upon by all the parties, that "at least one out of four representatives be a woman."[18] This inclusion was an important strategic victory for women, and it averted the potential disaster of almost complete exclusion from the talks.

Strategic concerns for women during this phase of negotiations have included ensuring a commitment to substantive equality in a future constitution; securing women's participation in areas that had previously excluded them, such as security and the economy; providing women's input on the "rules of the game" for the electoral process, including the establishment of an independent electoral commission, the development of a code of conduct, and the reintegration of the homelands; establishing a gender-sensitive, independent media; and, perhaps most important, lobbying for women's participation in the Transitional Executive Council (TEC) —the most powerful decisionmaking body of the transition.

Constitution-Writing Process

It is hoped that women will be well represented in the constituent assembly/new parliament. However, even with an exceptional performance at the polls, it is unlikely that women will be represented anywhere near in proportion to their percentage of the population, nor is there any reason to expect that all of the women elected will be concerned with the positions supported by the majority of South African women. Therefore, multiple approaches to ensuring the presence and visibility of women and women's issues will be a strategic concern during the constitution-making process.

Women have focused much of their efforts since the late 1980s on the constitution-writing process. A long process of debate and dialogue between the ANC leadership and the ANC Women's League has occurred over the ANC's draft bill of rights, and more recent drafts of the document have reflected women's recommendations.[19] The National Party has included provisions for women's rights in its draft Charter of Fundamental Rights. Progressive Western Cape lawyers have proposed substantive equality for women in a Charter for Social Justice. The purpose of the WNC's extensive participatory research effort is to articulate the needs and demands of women in a legal document for the constitution-making process, while educating women about their rights.

The participatory research undertaken by the WNC will provide several important functions with respect to the constitution-writing process: (1) it will attempt to identify and articulate the shared goals and priorities of a broad range of women; (2) it will produce a legal document, above the whims of political parties, that can be easily utilized by constitution-drafters; (3) it will educate women about the process underway to ensure their constitutional rights; and (4) it will provide an important research foundation to inform lawmakers in the development of the constitution and further legislation on women's issues. This research is expected to occur during mid-1993.

Substantive Equality Ensured in the Constitution

South African women have learned from the experiences of other countries that formal equality in a constitution is not enough to ensure equal treatment. They argue that formal equality must be supported by constitutional provisions that address the substantive issues of women's equality. This involves creating access to legal and constitutional remedies for inequalities that stem from particular disadvantages. Substantive equality seeks to "equalize outcomes" for men and women and to address the different realities of their lives.

Constitutional mechanisms will not completely ensure substantive equality. This process involves changing attitudes that are engrained in society and even reinforced by many constitutions, such as that of the United States, which feminists argue have inhibited substantive equality for women.[20] However, South African women have proposed several methods for enshrining substantive equality in a bill of rights/constitution. Different proposals for equality clauses have been written in such a manner to ensure equal outcomes through phrases like "women and men will be equal before the law" or "will have equal benefit of the law." Most of these proposals have clauses that prohibit discrimination based on sex, gender, or sexual preference.

Affirmative action and positive action clauses, as well as directives of state policy along the lines of the Namibian constitution, have been introduced to help remedy existing inequalities. Constitutions that provide for "gender equality," referring to the socialized behaviors of men and women as opposed to their physical attributes, are perhaps the most likely to succeed at eliminating existing social biases. The ANC has proposed a section on gender rights in its draft bill of rights, which prohibits discrimination based on "gender, single parenthood, legitimacy of birth or sexual orientation."[21]

Second, beyond the phrasing of an equality clause, the general language of a bill of rights/constitution has implications for substantive equality. The ANC has proposed the use of the terms "men and women" as much as possible to avoid ambiguities and ensure that women can "see themselves" in the constitution. Other proposals have argued that gender-neutral language is less confusing and avoids the problem of excluding certain groups by not listing them.

Third, a bill of rights/constitution that guarantees administrative review and judicial accessibility can help women use the constitution in favor of substantive equality.

Fourth, given that the majority of South African women experience discrimination mostly in the private spheres of the home and the family, a bill of rights/constitution that protects women from discrimination in these two vital areas will be better positioned to guarantee substantive equality. The current debate has centered on issues of vertical versus horizontal application, with the government on the side of the former, the ANC on the side of the latter, and other groups falling in between. In addition to whether the constitution is applied horizontally or vertically, clauses that protect the equal rights of partners in marriage, for example, can help protect women from discrimination in the home.

Issues of custom and religion form an important part of the public/private debate, and women have demanded that they be addressed in the constitution-making process. The ANC does not address customary and religious discrimination in its draft bill of rights, although the draft's horizontal application and other attributes appear to be more accessible to women faced with religious or cultural discrimination. The NP leaves the issue of customary law to the status quo, and its strict vertical application of a bill of rights excludes the private sphere.

Fifth, the most contentious debate to date has been over ensuring socioeconomic, or second generation, rights in a bill of rights/constitution. On one side are those who support a strictly liberal view of individual rights based on the U.S. Constitution as a model, such as the government's support for its draft Charter of Fundamental Rights. In the center are the authors of the Charter for Social Justice, who propose a negative constitution interpreted along the lines of a "free and open social democracy."[22] They argue that second generation rights cannot be enforced and have proposed Directives of State Policy to ensure that government policy includes a commitment to improving the socioeconomic position of the disadvantaged.

The ANC believes that second generation rights are enforceable in a bill of rights/constitution. Article 17 of its revised draft contains instructions to courts for the enforcement of second generation rights. Feminist critiques of constitutional proposals by ANC women argue even more strongly for positive enforcement of second generation rights. These women believe that rights are "indivisible," and it is therefore not necessary to distinguish between a civil right, such as the right to vote, and a social right, such as the right to housing. They reject the hierarchical notion of rights under which women's rights have been viewed as less important than human rights and social rights as less important than civil rights.[23] Many women believe that the enforcement of second generation rights will be one of the most important vehicles for improving the status of women, who are the most disadvantaged in terms of socioeconomic development.

Type of Constitution: The Federalism Debate

The extent to which federalism/regionalism is developed during this process will be critical for women, who must carefully consider how the powers delegated to regions affect their lives. For example, situations where states are given control over social welfare issues, where states make their own legislation on abortion, or where the constitution of a particular state or region provides for a greater jurisdiction of customary law could have a tremendous impact on women.

The constitution of the state of KwaZulu provides an interesting example in this regard. While the constitution would be considered discriminatory by many women because it proposes to "recognize and protect the application of traditional and customary rules" without reform, it actually provides a clause on "procreative freedom" that would permit women to "terminate an unwanted pregnancy when safe."[24] The implications of state powers in this case could be comparable to experiences in the United States and Canada. One envisions men moving to KwaZulu to enjoy the benefits of customary law, or women traveling to KwaZulu to obtain legal abortions!

Abortion

The issue of abortion in a future constitution has been continually raised by women only to be ignored by male decisionmakers. Abortion exposes race and class divisions in that rich white women have access to overseas abortions while poor black and rural women do not. There is a strong religious lobby that opposes abortion. Even progressive women's organizations like the Black Sash held a position against abortion for many years, until it adopted a pro-choice position at its 1993 national conference.

In their respective proposals, the NP leaves the issue to a future constitutional court and the ANC leaves the issue to legislation. The ANC position is a significant departure from the ANC Women's League position put forth at several forums in support of reproductive freedom.[25] Women across the political spectrum have criticized the major players for inadequately addressing the issue.

If the status of abortion is going to change from the current position of indifference by the negotiators to an informed position for a future constitution, women will have to exert a significant amount of muscle in the negotiations. If women identify the issue as a strategic priority, then abortion will be an important battle in the constitution-writing process.

CEDAW

The United Nations Convention on the Elimination of Discrimination Against Women (CEDAW) has played an important role in advancing the status of women worldwide. At the Lusaka conference in 1989, ANC women demanded the ratification of CEDAW by a democratic government and proposed that it be used as a guideline in the constitution-writing process. Progressive women's organizations began researching methods for the implementation of CEDAW in the South African context.

In January 1993, the de Klerk government signed CEDAW, sparking much controversy and fear that it hijacked an issue that had been a central pillar for women in the liberation movement. The government justified its rushed proposed legislation on women's equality on the grounds that "purging the books of discriminatory statutes" is a natural follow-up to signing CEDAW. Yet many women believe that by the government signing CEDAW, before black women could even vote, the importance of women's rights as human rights enshrined in CEDAW would be diminished. However, even NP women conceded that the government could not ratify CEDAW—it must be ratified by a future democratic government.[26]

However, despite the current tension over the issue, CEDAW received the attention and support of a broad spectrum of parties. In light of this, a commitment to the principles of CEDAW in the writing of the constitution, supplemented by a women's charter, would help ensure that the demands of women are accommodated and that the mechanisms to ensure their rights conform to international human rights principles.

Decisionmaking Structures for Women

Another strategic issue for women in the creation of a new constitution and a democratic government involves political decisionmaking structures for women. Women have already initiated the debate about the types of structures for women that could exist in a future government. A forum sponsored by the Institute for a Democratic Alternative in South Africa (IDASA) in December 1992 brought together women from organizations across the political spectrum to discuss structures for women in government. At the workshop, a consensus emerged that South African women should utilize a "package approach" of structures and mechanisms that would target different levels of government, the judiciary, political parties, and organizations of civil society. It was agreed that the package should be carefully linked to the constituencies of grassroots and rural women. Women also expressed the need for these structures to be given sufficient operating budgets. A package of women's decision-making mechanisms could provide a comprehensive and flexible approach to integrating women into a new democratic government and encouraging women's participation in public life.

Different political parties and organizations have proposed different mechanisms. There is consensus among many groups, including the ANC, DP, and NP, that a system of women's desks, integrated throughout government departments, should be developed along the lines of the system of Women's Units in Australia. There is also interest in an independent commission on the status of women, such as the Canadian Advisory Council on the Status of Women. Various parties have put forth proposals for different types of quotas or reserved positions for women, but there is little consensus about what types of quotas will best ensure women's participation in public life.

Transformation Issues: Reconstruction and Beyond

Beyond the constitution-making process and the establishment of South
Africa's first democratic government lies the challenge of consolidating
democracy and "delivering the goods" of democratic government to the
people. If a democratic transition is to lead to women's full participation
and equality in public and private life, then issues relating to the position
of women in the economic and cultural life of the country, as well as
women's physical safety both inside and outside the home, must be addressed. If democratic transformation is to succeed, the following issues,
at a minimum, must be addressed.

Legal Access for Women and Rights Education

Even if a commitment to substantive equality is achieved in the constitution, mechanisms must be created to give women affordable and timely access to law. Methods to claim child maintenance, access to magistrates,
and legal protection for rural women are just a few of the demands made
by women that are relevant to legal accessibility. Furthermore, if women
are not aware of their rights, they will not be able to invoke the constitution in their favor. The experience of Zimbabwe has shown that despite
some important legislative gains for women, improved access to courts,
and changes in the relationship between customary law and common law,
most women are not aware of their rights or the process by which they can
challenge the abuse of those rights through the courts.[27]

The logistical challenges of creating a system of accessible legal administration are daunting given the demographic realities of South African
women. Conferences on administrative law and judicial reform have discussed issues of access and administration and the overall restructuring of
the judiciary. There have been different proposals for equal opportunities
commissions and ombuds to deal with issues of sexual discrimination and
maladministration. Other options under consideration include the participation of lay people in the judicial process; special land claims courts and
other specialized tribunals; a revised role of traditional courts; and the expansion of the role of existing legal advice offices, such as those provided
by the Black Sash and Lawyers for Human Rights in many townships and
rural areas.

The experiences of different countries, such as Australia, have shown
that national publicity campaigns to promote awareness of the mechanisms
and institutions established to deal with legal administration have been effective in informing citizens of how their institutions can be utilized. Education about these institutions is essential if ordinary citizens are to use
them effectively. The cost of legal services will be an important consideration given the economic position of most South African women.

Customary Law Reform

Large numbers of South African women believe that issues of custom and religion as they affect women's lives have not been adequately addressed during transitional discussions. Yet these issues are most critical if the majority of women in South Africa are to enjoy full participation in a democratic society.

Women have identified customary law reform as a top priority. According to Mary Maboreke, "Most African people have neither completely moved away from the customary way of life, nor have they remained squarely rooted in it. In most cases they have a foot in each world."[28] Customary law intersects with women's modern lives in such areas as property ownership, inheritance, and child support. African women have sought to educate a broad spectrum of South Africans about the reality of customary law and its implications for their lives. The abolishment of polygamy, the initiation of a debate over *lobola,* and the recognition of African and Muslim marriages by civil law are frequently raised by women at various forums.

Despite the necessity for reform articulated by women, the NP has opted for a position that essentially maintains the status quo, although its recent pamphlets on women's equality promote the party's decision "not to interfere with the choice of women to live under the current system of indigenous law." Many women argue that in fact there is no choice in this issue at all—customary law is chosen for them by their husbands and their society. The ANC proposals say little about customary and religious discrimination against women, although ANC women have argued against discriminatory practices of traditional leaders during ANC discussions on the future role of traditional leaders and courts. The KwaZulu constitutional proposals "recognize and protect the application of traditional and customary rules not inconsistent with the provisions of [the] constitution," and further state that "traditional and customary rules . . . shall not be modified or repealed by the law."[29]

Although current transitional players have failed to address the issue, a debate is slowly beginning to develop among women, progressives, traditional institutions, and religious institutions regarding the future role of customary laws in a democratic society. Several forums have been planned for 1993 to engage women and traditional and religious institutions in a process of dialogue that will be crucial to evolving a position of custom and religion in accordance with the principles and laws of a democratic society.

Measures to Protect Women from Domestic Violence

Domestic violence and rape are of immediate concern to women across the political spectrum. Women's organizations such as Rape Crisis Cape Town and others have been effective in mobilizing around this issue. A recent initiative by lawyers, academics, social workers, and the attorney general in the Western Cape to form a special rape court is an example of one cooperative approach. Township-based organizations have also emerged to deal with rape and battery. South African universities have conducted inquiries into sexual harassment and abuse and have developed sexual harassment policies and counseling services. The marital rape exemption in the government's draft legislation on domestic violence caused an outcry from women of all political and social backgrounds, demonstrating that women demand a real commitment to this crucial issue.

Domestic violence is a sensitive issue for several reasons. First, it is not a clear-cut apartheid issue—members of the democratic movement are involved in domestic violence and rape. At the launch of the Western Cape Women's Charter Campaign, several woman articulated that "sexual harassment—*especially* at rallies and meetings—[must be addressed]. Women must be able to speak without fear."[30] While politically motivated violence has taken precedence over domestic violence, many women in South Africa argue that there is a strong political content to domestic violence. According to Mamphela Ramphele and Emil Boonzaier,

> The social and political order in South Africa impacts on working class men in a way that brings out the worst kinds of chauvinism in them. Black women present the only cushion against their complete powerlessness. . . . The oppression they suffer in the wider society acts as a paradigm for their domination of women.[31]

Furthermore, there appears to be a lack of recognition among the male peace negotiators that women are in fact victims of the political violence.

Many women's organizations have criticized constitutional proposals that fail to address the private spheres of the home and the workplace, where sexual harassment and physical abuse occur at alarming rates. Women have also demanded a restructured police force that is sensitive to the needs of rape and battery victims. The traditional police position of not interfering with domestic affairs must be transformed into one that accommodates victims and provides safety and protection.

Economic Empowerment

Much of women's work in South Africa is not recognized as formal labor. Even within the formal workforce, both the labor organizations of the democratic movement and the private business community continue to discriminate against women. These groups are the primary players in the National Economic Forum and other business groupings formed to lobby the transition. However, there has been significant progress by women in these areas, which, however slowly, may affect the long-term process of transformation.

In the business community many large companies, such as Pick & Pay and Southern Life, have developed gender equity policies and improved their capacity to accommodate the needs of women workers. A cooperative effort by the Women's Bureau and the University of Pretoria has utilized those companies with favorable gender policies to encourage similar practices in other companies. The National Manpower Commission has been restructured and now includes two women.

In late 1992, the International Labour Organisation (ILO) began consultations with a newly formed South African employers consultative group, consisting of the nine major employer organizations in the country. Nearly all of these organizations are male-dominated mining groups that employ very few women. Businesswomen have focused on finding avenues for women's participation in these negotiations, but have found it extremely difficult to convince these major employers to acknowledge the gender component in the economic transition.[32]

The gender debate rose in visibility in the trade union movement when, at the 1989 COSATU congress, women put two issues on the agenda: (1) the sexual conduct of shop stewards and (2) the lack of women in the leadership of the unions.[33] The congress debated a proposed resolution on these issues for more than four hours only to defeat it. However, the defeat served to mobilize women, and at the 1991 COSATU congress, women won the right to continue building women's forums and establish mixed-gender forums, to employ a full-time gender coordinator to oversee affirmative action programs, and to receive a proper allocation of resources for these activities.[34] Issues of child care and shared parenting are now on union agendas. There is optimism that over time women will be freed from serious issues of harassment and discrimination in order to rise appropriately in the labor movement.

Outside the formal labor and business spheres, many women are engaged in informal sector business. Debbie Budlender suggests that although the informal sector is often offered as a solution to problems of general subsistence and access to employment, most women enter the informal sector "out of desperation at the lack of alternatives. And even in this sector, women generally earn less than men."[35] Budlender argues that economic development policies to empower women must not only address their current situation, but also create a range of options for women in the economy.

Since the shift in the political climate brought on by the events of February 1990, women's organizations and NGOs have added issues of development and reconstruction to women's activities. Along with international trends, two approaches have been utilized in this process. The

Women in Development (WID) approach aims to integrate women into the development process. WID has gradually ushered in the Gender and Development (GAD) approach, which assumes women's integration and addresses both the strategic and practical needs of women in the development process. Organizations involved in this process include the Women's Health Policy Project at Wits University, the Rural Women's Movement, the Natal Organisation for Women's Empowerment, and the Women's College in Cape Town, among others.[36]

It is critical that these groups develop links with policymaking organs if women's development issues are to be determined a priority during the transition and beyond. The appointment of a gender coordinator by the National Land Committee is a good example of an attempt to establish such a link for women's land and housing concerns. The success of women in ensuring their inclusion in the negotiations over land policy is essential since, according to Budlender, "In all cases, the greatest problem facing all women is their lack of legitimate access to land. Land means not only a possibility of production, but also housing, shelter, security, and access to political power."[37]

Affirmative Action

Success of affirmative action in both the public and private spheres is the crux of women's political, economic, and social empowerment. If effective strategies are pursued, black women stand to gain the most from affirmative action programs. The gender/race balance of affirmative action will be an important consideration in the development of programs. Women's particular disadvantages must be considered based on their diverse experiences. Emphasis must be placed on the training components of these programs if women are to succeed in the long term. Programs must also seek to integrate women into a broad spectrum of job categories, as opposed to reinforcing their existing inequalities. Women must be trained to be scientists and doctors, not only nurses and teachers. Affirmative action must consider developments in technology as they relate to women's skills and not simply focus on small-scale income-generating projects.

If affirmative action is to succeed, evaluation mechanisms must be built into programs to continually examine their successes and failures. Women must participate in this evaluation process to ensure that their needs are met. The goals and criteria for programs must be transparent if they are to build the confidence of those affected by them. Special structures, such as equal opportunity commissions, must be given not only the autonomy to develop programs, but also the powers to enforce those programs. Several of South Africa's major universities have developed high-profile affirmative action programs and appointed gender coordinators to oversee them. These efforts provide an important learning experience for developing a comprehensive approach to affirmative action in South Africa.

Impediments to Transformation

There are many obstacles to women's success during the narrow transition and broader transformation. First, structural and contextual factors of the negotiations themselves will impede women in the transition. It is unlikely that women's issues will be considered a priority by most of the negotiators, policymakers, and constitution writers. Women will be accountable to the male leadership of their political parties. There is no representation for women as women in the process, although the GAC has been important in this regard. The security situation and prevailing atmosphere of violence are likely to remain threatening to the participation of large numbers of women in the process.

Furthermore, uncertainty as to the outcome of the transitional negotiations will make it difficult for women to develop long-term strategies. On the one hand, the deal cut by the ANC and the NP on the broad framework of the transition largely excluded women's concerns. On the other hand, within that framework, there is little agreement on details among the various parties. This will pose difficulties for women, even if the WNC Charter Campaign is successful at articulating the range of women's demands and bringing them to the table. Women must identify these structural obstacles and develop a flexible approach that both responds to the dynamic transitional environment and considers the consequences of transitional decisions in the context of a longer-term process of transformation.

Second, the most widespread impediment to women in both the narrow transition and the broader transformation process is patriarchy. ANC National Executive Committee member Albie Sachs has said, "It is a sad fact that one of the few profoundly non-racial institutions [in South Africa] is patriarchy."[38] Patriarchy is at the crux of discrimination and violence against women. It has determined the priorities of the transition process. It is found in nearly all of the structures and agents of transition and in every political organization.

The now famous quota debate at the forty-eighth ANC National Conference in Durban in July 1991 provides an important example of patriarchy within the liberation movement. According to Pat Horn,

> It had previously been accepted by the Constitutional Committee, the outgoing NEC, and all the Regions at an Inter-Regional workshop, that as an affirmative action programme to increase the participation of women in the ANC, there should be a minimum quota of 30 percent women of the 50 elected delegates.[39]

When the conference commission on the constitution proposed to drop the quota, a lively debate ensued. Most women argued in support of the quota. Of the mixed male reaction Horn reports that "some of the interventions were just crudely sexist," others were concerned with "the disappearance of merit as a criterion," and others supported the women.[40] After a long debate, the issue of the quota came to a vote.

> And, partly because of the lack of affirmative action in the ANC up to now, only 17 percent of the Conference delegates were women. So the tried and tested method of resolving issues by majority vote was in danger of perpetuating, rather than resolving, the problem. As a result after the vote was commenced, a member of the ANCWL delegation announced that the ANCWL would be abstaining from the vote. . . . The conference fell apart and women, supported by many men—including men and women who had voted against the quota, sang and demonstrated in favour of the quota.[41]

The very next proposal to give Umkhonto we Sizwe commanders automatic seats on the NEC received nearly unanimous support, without objections "on the basis of merit or democracy."[42] The quota debate illustrates that although the ANC has made significant progress in developing policy proposals designed to ensure women's equality, the organization has a long way to go before these policies will be practiced by the majority of its members.

The parliamentary debates over the NP's draft legislation on women's equality provide evidence for the existence of patriarchy within South Africa's predominantly white political parties. The following comment was made by a National Party MP in *support* of the legislation:

> While women do not want to forfeit their femininity, . . . it is also necessary that woman as a worker and an achiever should be viewed in a neutral manner. Sexist remarks, often well-intentioned, can be experienced by women as very degrading. While a compliment on her appearance is appreciated, she also asks for recognition for her contribution to other fields.[43]

The clearly sexist jokes and comments made by members of the Conservative Party during these debates further attest to the existence of patriarchy in the white male culture of South Africa. The influence of the all-male Afrikaner Broederbond on white politics is perhaps the most important example of patriarchy in the Afrikaner political culture.

A third impediment to women's success in the transition and transformation process is found in differences among women themselves. Many women point to the lack of a strong women's movement in South Africa. Rhoda Kadalie argues that because political activities of both white and black women in South Africa have been based on race consciousness and not gender consciousness, such a movement has not emerged.[44] The WNC has been the broadest attempt to create a women's movement. However, anger over the WNC's limited mandate and the use of the WNC issues by the NP, while it unilaterally proceeded with its draft legislation, have threatened to destroy the WNC.

The WNC has continued to exist by agreeing to work across differences. However, the historical experiences of South African women will continue to test the fragile coalition. Kadalie suggests

> that black women initially articulated their concerns primarily in terms of racial oppression, relegating their oppression as women to the background. . . . [This] was to silence the very important voices which [are] crucial to our understanding of real democratic transformation. This is the fulcrum around which such a coalition will evolve.[45]

Whether through the current form of the WNC or a more loosely co-ordinated women's effort, women must maintain pressure on the process and provide information about their issues and concerns. The patriarchy entrenched in South Africa's political leadership will not disappear during the process. If violence accelerates, opportunities for women will narrow. Thus, women must adopt a flexible strategy that considers the consequences of certain priorities and not only utilizes the multiple institutions of political parties and advocacy groups, but also works as part of a coordinated national women's effort. This strategy must seek to incorporate the politics of difference that have tested the WNC.

Notes

1. *Wathint' Abafazi: Stories of Four Women Who Were Beaten by Their Husbands* (Cape Town: Battered Women's Working Group, 1992).

2. Quoted in the *Star,* March 23, 1993.

3. Cherryl Walker, *Women and Resistance in South Africa* (Claremont: David Philip, 1991), pp. xiii–xiv.

4. In this section, the description of women in South Africa is based on preliminary work of the Women's National Coalition research coordinator, Debbie Budlender, and from an unpublished report by Shireen Hassim, Centre for Social Development Studies, University of Natal, Durban. Most statistics on South African women fail to truly reflect the reality of their situation. First, official statistics exclude the "independent homelands" of Transkei, Bophuthatswana, Venda, and Ciskei (TBVC areas), where large numbers of women live. Second, few socioeconomic statistics in South Africa are disaggregated by gender at all. For example, the *1991/1992 Race Relations Survey,* considered a comprehensive statistical yearbook, mentions women on fewer than twenty of its more than 600 pages and provides only one statistical table disaggregated by gender. Third, statistics

often consider women broadly and do not examine the wide variances among
women of different races and economic backgrounds.

5. Statement by Carole Charlewood, MP, Democratic Party, in *Debates of
Parliament: Fifth Session—Ninth Parliament, 8–12 February 1993* (Cape Town:
Government Printer, 1993), p. 1160.

6. Cathi Albertyn, "You and Your Marriage," *Speak*, March 1993, p. 23. The
government has recently proposed legislation to abolish marital power. This legislation was to be debated in mid-1993.

7. Information on customary marriages was taken from Cathi Albertyn, "You
and Your Marriage: Customary Marriages," *Speak*, April 1993, pp. 14–15. Customary marriages are not recognized by civil law, only by customary law.

8. Crecentia Mofokeng and Thembi Tshabalala, "NACTU Women Speak
Out," *South African Labour Bulletin* 17, no. 1, p. 72.

9. Jacklyn Cock, *Maids and Madams: Domestic Workers Under Apartheid*
(London: The Women's Press, 1989), p. 7.

10. Desiree Hansson, "Working Against Violence Against Women: Recommendations from Rape Crisis," in Susan Bazilli, ed., *Putting Women on the Agenda*
(Braamfontein: Ravan Press, 1991), p. 181.

11. *Violence in South Africa: The Report of the Commonwealth Observer Mission to South Africa*, Phase I: October 1992–January 1993, p. 12.

12. Bazilli, *Putting Women on the Agenda,* p. 18. Hansson reports that in 1985
alone, 32,500 women (of whom 13,600 were between fifteen and nineteen years
old) had surgery to complete botched illegal abortions. See Hansson, "Working
Against Violence," p. 189.

13. Navi Pillay, "Judges and Gender," in *Agenda: A Journal About Women
and Gender*, (Durban), no. 15 (1992), p. 47.

14. *Agenda*, no. 13 (1992), p. 23; "Press Release of Gender Advisory Committee," March 16, 1993. The GAC was made up of one member of each organization represented at Codesa, and its terms of reference included advising the management committee on gender issues and scrutinizing the decisions of Codesa in
terms of their gender implications.

15. Pregs Govender, unpublished paper, Women's National Coalition, February 1993.

16. National Council Meeting Report to the Western Cape Branch of the
WNC, February 1993.

17. National Conference: Women and a Charter of Fundamental Rights,
March 8, 1983, Pretoria.

18. *Weekend Argus*, April 3–4, 1993.

19. See Dorothy Driver, "The ANC Constitutional Guidelines in Process: A
Feminist Reading," and articles by Murray et al., Ginwala, Mabandla, and Hansson
in Bazilli, *Putting Women on the Agenda.*

20. See Melissa Haussman, "The Personal Is Constitutional: Feminist Struggles for Equality Rights in the United States and Canada," in Jill M. Bystydzien-
ski, ed., *Women Transforming Politics: Worldwide Strategies for Empowerment*
(Bloomington: Indiana University Press, 1992), p. 70.

21. *ANC Draft Bill of Rights* (preliminary revised version), February 1993,
p. 10.

22. Hugh Corder, et al. *A Charter for Social Justice* (Cape Town: Legal Resources Centre, 1992), p. 29.

23. Conversations with ANC women and meetings of the National Association of Democratic Lawyers, January to April, 1993.

24. *The Constitution of the State of KwaZulu/Natal*, December 1993.

25. Bridgette Mabandla in Bazilli, *Putting Women on the Agenda,* p. 77. Mabandla reports that demands made by women at the 1989 ANC conference in Lusaka included that "the law should enable women to have full control over their bodies and that they should have the right to choose to have a child or terminate a pregnancy."

26. Interview with Sheila Camerer, March 1993. Even within the NP, particularly among the men, there is significant confusion as to whether the government has signed or actually ratified CEDAW. Camerer confirms that it has only signed the convention.

27. Mary Maboreke, "Women and Law in Post-Independence Zimbabwe: Experiences and Lessons," in Bazilli, *Putting Women on the Agenda.*

28. Ibid. p. 220. Maboreke describes how Zimbabwe has attempted to reform its customary law system through the review of choice of law criteria between common law and customary law. This experience has had important implications for women.

29. *The Constitution of the State of KwaZulu/Natal,* December 1992.

30. Problems and Demands of the Women's Charter Campaign in the Western Cape, August 1992.

31. M. Ramphele and E. Boonzaier, "The Positioning of African Women: Race and Gender in South Africa," in E. Boonzaier and J. Sharp, eds. *South African Keywords: The Uses and Abuses of Political Concepts* (Cape Town: David Philip, 1988), p. 166.

32. Interview with Professor Ronel Erwee, University of Pretoria, March 1993.

33. Rhoda Khadalie, "Women's Organisations in South Africa: Models for Peaceful Co-Existence in South Africa," unpublished paper, p. 20.

34. Fiona Dove, "Clearing the Gender Hurdles," *South African Labour Bulletin* no. 8, p. 67.

35. Debbie Budlender, "Women in Economic Development," in Glenn Moss and Ingrid Obery, eds., *South Africa Review 6: From 'Red Friday' to CODESA* (Johannesburg: Ravan Press, 1992), p. 359.

36. Michelle Friedman, "Fine-tuning the Gender Agenda," *Work in Progress* (February 1993). Friedman defines women's practical needs as those that arise out of the customary division of labor and are of daily concern, such as water, shelter, and health care. Women's strategic needs address issues of gender relations such as political power and access to employment. Strategic needs must be addressed over the longer term.

37. Debbie Budlender, "Rural Women: The 'Also-Rans' in the Development Stakes," *Agenda,* no. 12 (1992).

38. Bazilli, *Putting Women on the Agenda,* p. 9.

39. Pat Horn, "ANC 48th National Conference," *Agenda,* no. 11 (1991), pp. 15–16.

40. Ibid., p. 16.

41. Ibid., pp. 16–17.

42. Ibid., p. 117.

43. *Parliamentary Debates,* p. 1175.

44. Kadalie, "Women's Organisations in South Africa," p. 7.

45. Ibid., p. 27.

7

Race Relations Law in a Reformed South Africa

Thomas G. Krattenmaker

The fundamental principles of South African law regarding race relations are undergoing substantial change. In particular, South African law now appears to accept, or to be on the verge of accepting, what I will call an "anti–race discrimination principle." This principle forbids statutes or rules employing explicit racial classifications that are intended to subjugate, segregate, or otherwise disadvantage a disfavored racial group. And there is some reason to hope that South African law may soon incorporate a principle that forbids racial discrimination in assigning the right to vote.[1]

In this chapter, I will not try to guess precisely when or to what extent these revolutions in South African law will occur. Rather, assuming (or hoping) that they will occur, I will attempt to define and categorize the subsequent legal and political controversies over the precise contours of South African race relations law that are virtually certain to arise on the heels of South Africa's adoption of the anti–race discrimination principle and the principle that voting rights should be assigned without regard to race. In this fashion, I hope to provide a preview to some of the race relations issues one can expect to see debated in South African law for the next several years.

Legal developments in the United States and elsewhere teach that being rid of overt racially discriminatory legislation and granting the franchise without express racial qualifications are only the first, albeit the most important, steps in moving toward a nonracial democratic society.[2] The issues addressed here arise whether these initial fundamental principles are adopted by a parliament or a constitutional convention and whether they are enforced by the executive or the judiciary.[3]

In the first part of this chapter, I argue that a prohibition on statutes or rules that employ explicit racial classifications that are intended to subjugate, segregate, or otherwise disadvantage a disfavored or vilified race (U.S. "Jim Crow" laws) will not end the controversy over what the basic contours of race relations law should be in a nonracial society. Rather, the society that abolishes overt, punitive, racially discriminatory legislation will almost certainly and rather quickly find itself confronting three other questions about the contours of its race relations law: the scope of permissible, private discrimination; the legal status of unintentional discrimination; and the legitimacy of race-based remedial legislation (or affirmative action).

In the second part of the chapter, I argue that adopting the principle that the right to vote shall be assigned without regard to race will confront South Africa with a subsequent serious dilemma: whether to embody that principle in a law that is easily understood and applied but potentially ineffective or, rather, to adopt a legal rule that appears more sophisticated but may prove incoherent or useless in some cases.

Elaborating the Contours
of the Anti–Race Discrimination Principle

As noted above, South African law has not abolished racial discrimination, although it has taken relatively large strides in that direction. For present purposes, however, assume that the anti–race discrimination principle described above is enshrined in South African law. What issues, previously submerged in a legal regime that fostered discrimination, will then emerge?

The Problem of Private Discrimination

No democracy, no matter how committed to achieving equality, can outlaw all forms of racial discrimination. A race-based refusal to permit someone to join the neighborhood card game or the failure to include people of color on one's wedding invitation cannot be outlawed without first establishing an intolerably intrusive police state.[4]

On the other hand, if the legal rule that emerges from commitment to the anti–race discrimination principle is that only the government is bound to refrain from discrimination, promises of liberation may indeed ring hollow. May people of color be excluded from universities, from high-ranking jobs in major corporations, from the country's great shopping centers, or from large housing developments by the nongovernmental owners of these institutions? If so, one wonders how members of disadvantaged races may ever be able fully and effectively to participate in political aspects of society if they are excluded from schooling, jobs, goods, and housing that make possible daily enjoyment of the nonpolitical benefits of that society. May government avoid its obligations by narrowing its responsibilities, selling schools or swimming pools or public corporations to private entities that can then discriminate?[5] If so, then the areas within which freedom from racial discrimination may be enjoyed could prove to be very narrow indeed.

As these rather simple hypothetical examples show, it is both necessary and difficult, in a society committed to humane race relations law, to mark out the proper respective bounds of public and private decisionmaking spheres. The task is difficult because it involves balancing the commitment to equal opportunity and participation with the concomitant quest for a democratic society free of excessive intrusion in private matters, while also attempting to determine what are the essential roles of the state that may not be delegated to nongovernmental bodies. Obviously, different societies and different cultures will strike the balances among these fundamental values differently. But, at a minimum, calls for effective controls over private discrimination in such key areas as housing, employment, schools, and access to places of public accommodation will be persistent and earnest in a society moving toward a nonracial democracy.

The problem, as I see it, is somewhat more acute in South Africa than it is in the United States because of differences in the two countries' economic organizations. A truly open, competitive, and diverse private sector can mitigate the need for legal regulation of discrimination within that sector. Discrimination may not be as consequential or as durable if the disfavored group enjoys equally attractive alternative nondiscriminatory outlets.[6]

It is easier to speak of a multifaceted, competitive, and accessible private sector in the United States than it is in South Africa. While portraying itself as a capitalistic society, built on individual initiative and free enterprise, the government white South Africans established in fact has practiced a form of state socialism for at least the past four decades.[7] Government is so incestuously intertwined with industry and higher education, for example, that it is quite difficult to argue that a proper respect for private zones of decisionmaking compels a narrow scope for the anti–race discrimination principle. Most major industries in South Africa appear to be protected from competition by government limitations on entry. Universities are not free in South Africa; they are established only by permission of the government, and, until recently, each was rigidly confined to admitting people of only one race. Thus, for example, the argument is not very compelling that one could permit the mining company or sugar refinery or medical school to hire only whites on the grounds that these are only the voluntary decisions of nongovernmental units whose competitors may behave differently.

The Problem of Unintentional Discrimination

Citizens in a nonracial South African state will repeatedly have to answer the question "What is the discrimination we seek to eradicate or remedy?" Of course, subject to resolution of the public-private boundary dispute just discussed, unlawful discrimination would include rules or standards that explicitly employ race with the purpose of subjugating or segregating people of color. But what of rules that bear more harshly on disadvantaged races although they were not expressly designed for that purpose? Should these rules also be judged "discriminatory"?

A leading case from the reports of the Supreme Court of the United States, *Washington v. Davis,*[8] neatly illustrates the problem. Local authorities in the District of Columbia administered a test of general knowledge and communicative skills to all applicants for jobs on the city's police force. No one ever tested the test, i.e., analyzed whether in fact those who passed made better police officers than those who did not. It was simply assumed, not illogically, that better educated, more capable communicators would make better police officers.[9] But someone did test the test to see how it affected applicants. This analysis showed that a much higher percentage of black than of white applicants failed the test. The city was unable to cite any factor other than race that appeared to account for the differences in performance.

Was this racial discrimination? From the standpoint of a job applicant, it must certainly have appeared to be so. Because of the government's regulation, if you were a black person, your chance of obtaining a job was substantially less than if you were white. The District of Columbia might just as well have enacted a statute providing that, in all police hirings, white job seekers were to be preferred over black applicants.

But the U.S. Supreme Court concluded that utilizing the test did not constitute racial discrimination in the constitutional sense because no one intended to distinguish among applicants on the basis of race. The disproportionate impact on black citizens was "merely" an unintentional by-product of a perfectly legitimate, purposeful, and explicit discrimination against people who got lower test scores. Administrators could (as in this instance they later did) choose not to administer the test because of its differential effects, but the anti–race discrimination principle did not require them to do so.

Why would one wish to condemn only intentional racial discrimination in a just society? One way to understand the Supreme Court's rationale is to consider what would be required had the Court resolved the matter differently.[10] If the disparate impact of the employment test had rendered its use illegal, then it would appear that those who hired police officers would be required to act on overtly racial grounds. They would

have to either (1) not give the test to black applicants or (2) set different
passing rates for blacks and whites. That is, the principle of nondiscrimination would be turned into a legally binding requirement that government
must equalize outcomes on a racial basis (except in situations where members of disfavored races did better than members of privileged races) even
to the extent of expressly employing race as a legal criterion.[11]

This rationale for the *Washington* case, while impressive, is not necessarily convincing. Others might well assert that, in a society still emerging from a past of oppressive discrimination, unintentional discrimination,
in reality, is often or even usually a by-product of previous hostile discrimination.[12] For example, is it not likely that the results of the District of
Columbia's test were rooted in an inferior, segregated education that many
black Americans received before the fundamental law was altered to for-
bid racial segregation in public facilities?

Indeed, the U.S. Supreme Court has also been attracted by reasoning
like this. The *Washington* case involved the question whether the city of
Washington, D.C., had engaged in unconstitutional discrimination. In an
earlier case, *Griggs v. Duke Power Co.*,[13] the Court considered what Congress meant when it forbade private employers to deprive anyone of employment opportunities because of race.

In *Griggs*, an employer adopted a policy that required job applicants
(and current employees seeking to transfer to another line of work within
the firm) to have a high school degree or to pass two standardized general
intelligence tests. Of people living in the employer's area, white workers
were much more likely than black workers to have graduated from high
school and to pass the tests. Since many people who had been hired before
these requirements were adopted nevertheless performed admirably, there
was no serious contention that the requirements measured ability to do the
job. But lower courts had found the practices lawful because the employer
had no intention, in adopting the requirements, to discriminate on the basis
of race.

The Supreme Court held that the employer had committed racial discrimination within the meaning of the statute because the company had
adopted an employment practice that excluded black workers and that was
not shown to be related to job performance. The graduation and test requirements were "fair in form, but discriminatory in operation."

Washington and *Griggs* are technically consistent with each other because the former construes the Constitution while the latter reads a statute.
In fact, however, they represent the Court looking each way at the trou-
bling issue of unintentional discrimination. Adherents of the *Washington*
Court's view of discrimination would ask whether the *Griggs* Court was
sufficiently sensitive to the overt racial analysis that *Griggs* compels. Employers subject to the *Griggs* rule must not discriminate, but they must

constantly be acutely race conscious as they monitor the impact of all their policies on various racial groups. Proponents of the *Griggs* view of discrimination would argue that the *Washington* Court was insensitive to the underlying causes of the test's disparate impacts and disregarded the intermediate alternative of simply requiring that job tests be job validated.[14]

South African lawmakers and jurists will find the issues no less perplexing. As with the problem of private racial discrimination, the problem of devising a wise legal response to unintentional racial discrimination will be especially troublesome in South Africa. The chance that a society's laws will produce disproportionate impacts along racial lines is substantially affected by the extent to which benefits and burdens in that society are disproportionally distributed by race. Given the extreme income and wealth disparities in South Africa,[15] disparities that are possibly the world's greatest and that certainly rest on racial lines, this means that the effects of a myriad of laws in South Africa are likely to be felt closely along racial lines. Any statute that affects people differently depending on their wealth, income, education, job status, residence, or language—to cite a few examples—is almost certain to produce effects that correlate very closely with the race of the person affected by the law. And, of course, policymakers will not be able to dismiss easily the argument that they must, because of previous governments' responsibilities for these conditions, treat such unintended effects as unlawful discrimination.

The question of whether to treat unintentional discrimination as no less odious than intentional discrimination is in fact a question about the extent to which one believes that the ills of the past continue to infect the present society, as well as a question about the dangers of engendering race-conscious decisionmaking as an antidote to the race-explicit regime of the past. In this manner, the issue of unintentional discrimination is tied to the question of remedial or affirmative action.

The Problem of Affirmative Action

Suppose that, in granting access to educational or employment opportunities, people of color are explicitly favored. For example, a rule might provide that, whatever other criteria are employed, twenty-five seats in an entering university class will be reserved for black students[16] or that 50 percent of all new job openings in a certain category will be filled by black workers.[17] Would such rules be condemned by the anti–race discrimination principle?

To answer this question requires that one first answer another: Precisely what is it that was most objectionable about the previous system of racial discrimination? If the fault was that racial classifications were employed by privileged elites to dominate, subjugate, and impoverish members of disfavored racial groups, then this "reverse discrimination" is not objectionable at all, for it seeks to liberate, empower, and enrich those who suffered from previous harmful discrimination. But if the flaw in the previous system was that it judged people according to the color of their skin rather than according to their individual merit, that it engendered a political racial spoils system, that it reinforced habits and attitudes of racial stereotyping, then such discrimination is no better than the disease for which it is said to be a cure.

It is easy to overstate the significance of this issue. In many, perhaps most, cases remedial action need not employ any form of racial discrimination. A rule that equalizes per pupil primary school expenditures across the board does not discriminate on the basis of race.[18] A program that gives special management training to corporate employees who have not previously been given executive responsibilities does not discriminate on the basis of race. It is not racial discrimination to decide to spend less on defense and more on shelter for the homeless. Such laws and policies do not discriminate on racial grounds. And, although they may in practice largely benefit people of one race, the beneficiaries are not defined by their color but rather by their need or by the harm they have suffered in the past.[19]

Typically, it is only when race-neutral remedial action fails to bring about substantial equality in practice that demands for race-specific affirmative action are urged. As indicated above, such policies will be advocated on the grounds that they merely seek to correct the imbalances created by previous discrimination and attacked on the grounds that they perpetuate racial stereotyping by assuming that all members of formerly disfavored races were equally harmed by the previous laws and that all members of the privileged race were equally to blame for their enactment and enforcement.

In the United States, courts have often been reluctant to scrutinize carefully race-specific affirmative action on the grounds that majority white legislatures are unlikely to impose such disabilities on white people unless they are clearly called for.[20] If, in the new South Africa, the lawmaking body is composed predominantly of people of color, it will be less easy to conclude that the simple passage of a race-specific remedial measure proves its justification. On the other hand, legislators in a new South Africa may not need to resort to explicit racial classifications to ameliorate present evils. Simply dispersing government benefits without regard to race, returning property that was earlier confiscated in a racially discriminatory manner, and enforcing fair employment laws may go a long way toward redressing the current extreme disparities among "White," "Coloured," "Indian," and "Black" that now dominate the demography of South Africa.

Some Conclusions

Experience shows that in a society that has practiced pervasive racial discrimination, outlawing state-enforced segregation is an important and indispensable beginning in restructuring race relations law. It is, however, only a beginning and not an end. The society will rapidly confront a variety of subsequent issues that are quite vexatious, at least at the margin. Particularly troublesome issues are the extent to which private discrimination will be tolerated, the proper response to unintentional discrimination, and the permissibility of certain forms of remedial (or affirmative) action.

Issues such as these will move rapidly to the forefront of political and juridical debate over race relations law once the fundamental anti–race discrimination principle is firmly enshrined in South African law. Further, the unique characteristics of South African race relations history ensures that no ready answers to these questions will be found in the law books of other countries. In large measure, a new South Africa must devise its own response to the challenge to establish a humane race relations law for all its citizens.

Elaborating the Contours of a Prohibition on Racial Discrimination in Voting Rights

Black people of African descent constitute an overwhelming majority of South African citizens. However, black people have never been given any voting rights in South Africa (except for rights to participate in rather meaningless elections within the so-called homelands). A central tenet of those advocating constitutional reform in South Africa is that racial discrimination in assigning the right to vote should be prohibited.[21] Surely such a step is an indispensable prerequisite to achieving a durable democratic social order in South Africa.

Abolishing voting discrimination should also aid in protecting against other forms of racial discrimination in South Africa. Black, as well as so-called coloured and Indian, citizens have been more vulnerable to public discrimination because they have been denied meaningful participation in the conduct of South African governmental affairs. If the franchise were fairly distributed in South Africa, reformers might feel more comfortable leaving some thorny issues of race relations law—such as the permissible bounds of private or unintentional discrimination, discussed above—to legislative, rather than constitutional or judicial, resolution.

More than 120 years ago, the United States adopted a constitutional amendment forbidding race discrimination in voting.[22] Many statutory prohibitions have been added in the past four decades.[23] If the U.S. experience teaches anything, it is that outlawing race discrimination in awarding the franchise is not a simple matter or one that requires resolving only one issue. Thus, I seek to draw on the U.S. experience to illuminate some of the fundamental issues of constitutional law and of democratic politics that are likely to be debated and discussed in South Africa should that country take the noble, but complicated, step of seeking to define and outlaw racial discrimination in assigning the right to vote.

To those unfamiliar with this area of law, the task might seem quite simple. To end discrimination in voting, one would just provide, by constitution or by statute, that in every electoral system established in the country, the vote of everyone shall be counted equally, irrespective of race. Obviously, such a rule would make a large difference in the conduct of politics in South Africa. But whether it would be adequate to protect against discrimination in awarding the franchise should be a debatable issue. People will fiercely debate who is covered by the prohibition and precisely what constitutes voting discrimination.

Who Is Forbidden to Discriminate?

The issue addressed here is precisely the same as that raised above concerning private discrimination generally. What if a political party, not an arm of the state, adopts a rule that forbids members of a certain racial group from participating in its processes for selecting candidates to run in general elections?[24] If such a practice is not forbidden, a powerful party may effectively exclude that racial group from participating in South Africa's governance.

For example, suppose that under a new constitution in which all South Africans are entitled to vote many political parties arise. Suppose further that in one town, which remains majority white, the National Party adopts a rule that only whites can be nominated as NP candidates for the city council. If white voters elect only NP candidates, people of color will have little or no say in the city's governance even though they are fully entitled to vote on election days. Or suppose the African National Congress becomes the politically dominant national party but refuses to permit Indians to sit on the party's executive council. To the extent that the executive council affects the positions taken by elected ANC candidates, such an action would diminish the value of the franchise to South African citizens of Indian descent.

The simplest way to deal with such potential problems, of course, is to outlaw racial discrimination by political parties. But a simple rule that all political organizations must be nondiscriminatory may unduly intrude into individuals' political rights. Would a women's caucus within the National

Party be required to admit males? If a South African political action organization were formed to address issues, created by previous discrimination, of particular concern to persons formerly classified as coloured, would the organization be required to admit white and black voters—even if the numerical strength of the latter groups threatened to overwhelm control of the new organization by coloured citizens?

The problem, of course, is that the nondiscrimination principle, particularly when applied to voting rights, can be in tension with other core values, especially the rights to self-expression and self-determination.[25] In a true nonracial democracy, political parties would take pride in being utterly free of discrimination and would be punished at the polls when they were not. But, because politics in South Africa have always been expressly racially polarized, it will not be easy to assert that a broad prohibition on racial discrimination by private political organizations is appropriate. At a minimum, some of those who have suffered in the past from severe political discrimination may strongly believe that they have at least a temporary urgent need to establish political organizations that they control. I know of no simple way to define when such control transcends a necessary modicum of self-protection and becomes an effective denial of equality at the polling place.

What Is Discrimination?

The manner in which representatives are chosen also can substantially affect the value of the votes cast by members of different racial groups. Thus, a principle that there shall be no racial discrimination in voting may be ineffective or inadequate if not combined with control over the method of representation. However, to make matters more complicated, there is no system of representation that does not threaten some measure of discrimination when compared to its alternatives. The problems that arise in scrutinizing systems of representation are quite similar to those, discussed in the first part of this chapter, that arise in attempting to determine to what extent one should condemn laws as discriminatory when they affect people of different races differently.

I think these points are easiest to illustrate if we assume that a large city in South Africa wishes to be governed, in all or some matters, by a city council, consisting of ten elected council members. Two basic questions need to be answered in resolving how best to choose ten people to represent residents of the city. First, will each member of the council be elected from one of ten separate districts within the city ("single-member districts") or will all members be elected by all the people within the city ("at-large" elections)?[26] Second, how will winners be determined in contests under either system? Under a scheme of proportional representation, electors vote for parties and each party is allocated the proportion of council seats that correspond to the proportion of votes the party received. Alternatively, in a winner-take-all system, people may vote for candidates rather than parties. The candidate receiving the most votes gets the seat in single-member district cases and the top ten vote getters become council members in an at-large system.

What if the city decides to select the members of its city council by dividing the city into ten districts and selecting one council member from each district? Unless housing patterns within the city are completely desegregated racially, choosing single-member districts rather than selecting council members at large is quite likely to affect the racial composition of the council.[27]

For a simple example, assume that 90 percent of the city's voters are black while the remaining 10 percent are white, and that the white residents live in an isolated enclave. Suppose further that, to ensure some continuity on the council, council members will serve for two-year terms, with half the council standing for election each year. Under a single-member district system, one white person will probably be elected to the council every other year from the white enclave. If council members were to be elected on an at-large basis, however, it is highly probable that all of them would be black people.[28]

May racial groups (in this case, black voters) that would have done better under an at-large system complain that the single-member system constitutes racial discrimination because it dilutes their voting power? May racial groups (in this case, white voters) that benefit from single-member districting assert that they are entitled to that system because any other scheme would dilute their voting power? One cannot agree with both of these arguments because then no system of representation could be adopted. But if both arguments are rejected, it is permissible, but not required, to choose either system; then voters are free to choose the system that benefits their racial group at the expense of others. That is, in this case, the black majority could choose an at-large system in order to enhance black political power without violating the prohibition on racial discrimination in assigning voting rights.

Suppose the city is allowed to employ single-member districts. The delineation of these districts can raise equally intractable problems.

The principle of one-person-one-vote presumably requires that each district contain an equal number of voters (in order to ensure that citizens in each district have an equal say in the selection of city council members). So the city needs to be divided into ten districts, each with an equal number of voters. Suppose further that, in this city, people of Indian descent constitute a racial minority. Finally, suppose that in dividing part of the city into two districts of equal size there are basically two options.

Under option one, Indians will constitute a majority of voters within one district and a very small minority in the other. Under option two, Indians will constitute a significant minority in each district.[29]

Does choosing either option one (which virtually guarantees people of Indian descent one representative on the city council, but leaves other Indians virtually politically powerless in one district) or option two (which probably means no Indians on the city council, but may well mean that the representative in each district will be attuned to Indians' interests and concerns) constitute racial discrimination in voting? Do *both* options violate the nondiscrimination principle?

It might appear that the city could avoid the previous problems by adopting two principles: (1) all council seats would be filled by persons elected at large and (2) elections would follow the rule of proportional representation. In the first example, if voters cast their ballots along racial lines, a white candidate is likely to get 10 percent of the vote and thus one of the ten council seats would go to that person. In the second example, following the same reasoning, Indians would be able to obtain a percentage of council seats that corresponds to their percentage of the entire voting populace.

But, of course, this response does not avoid the charge of discrimination at all. If black (or white, or whatever) voters would have had *more* power under a single-member system or a system of voting for candidates rather than parties, then at-large proportional representation constitutes a dilution of that power and, for that reason, can be said to be discriminatory.

Indeed, the simple choice between proportional representation and winner-take-all schemes, in which voters choose among candidates rather than parties, can have racial effects. Consider again the first example, in which ten council members must be chosen from a city that is 90 percent black and 10 percent white, but assume the racial distribution of voters is the same in each voting district. Under a proportional representation scheme, a white candidate could receive 10 percent of the vote in each district and thus be awarded a seat. But if one council member is selected on a winner-take-all basis from each district, all council members are likely to be black. In a democracy that does not permit racial discrimination in voting, is the "correct" outcome that 90 percent of council seats should go to black people because blacks constitute 90 percent of voters (proportional representation), or is it that 100 percent of seats should go to blacks because blacks win 100 percent of the district elections (winner-take-all)?

In these examples, I have used city council governing bodies for illustrative purposes because I think the consequences are easier to grasp. It should be clear, however, that the difficulties apply equally at the level of the national legislature. One can distribute representatives to the national legislature either by districtwide (single-member) or nationwide (at-large) voting. Citizens in either type of election can vote for candidates (largest vote getter wins) or for parties (proportional representation). The problems with identifying what one means by a voting system for a national legislature that is free of racial discrimination are identical to those portrayed with respect to city councils.

Some Conclusions

I do not wish to be misunderstood. The foregoing is not an argument for maintaining the status quo in South Africa. Certainly, a cornerstone of constitutional reform for South Africa should be adoption of the simple principle that there should be no racial discrimination in assigning the right to vote. Such a principle is widely, perhaps universally, understood as an indispensable element of a truly democratic society. Without doubt, adoption of such a principle would very substantially increase the political power wielded by people of color in South Africa.

Reality teaches, however, that the simplicity of this principle can be deceptive. In truth, the principle of nonracial voting laws masks a serious dilemma that must be confronted at some time, although not necessarily at the outset of reform. To express that principle as a rule of law, one must apparently choose between a rule that is easily understood and easily applied, but potentially ineffective in some cases, and a rule that promises to have more teeth but is likely to prove incoherent or useless in some cases.

On the one hand, the nondiscrimination principle is simple to understand and apply if all it means is that, once a system of representation is set up, the vote of everyone counts equally regardless of race. If that is all the nondiscrimination principle means, however, the effective voting power of some racial groups may be diluted by private political organizations or by the choices made among various possible systems of representation.

On the other hand, the nondiscrimination principle has no readily ascertainable and acceptable meaning if it is taken to require that all citizens shall enjoy equal voting power regardless of their race. The goal of prohibiting racially discriminatory franchises may be thwarted if the law does not also reach private political activity that is racially segregated. Yet such a law might unduly interfere with the constitutional political rights of freedom of speech and association that are established to permit people to seek to define and further their own welfare. Moreover, the choice of geographical units from which representatives will be elected can dramatically affect the political power of various racial groups. This "problem" is not "solved" but only repackaged by expanding or contracting the geographical unit or by adopting or rejecting a system of proportional representation.

Ironically, it appears that there is only one, obviously unacceptable, way to guarantee complete racial equality in the exercise of the franchise.

That is to decree that (1) in every body of elected representatives each race shall have the same proportion of members of the body as its proportion in the electorate, and (2) voters may vote only for members of their own race. That such a response would be monstrously perverse[30] in a South Africa that is supposed to be turning away from a past of strict political segregation does not mean there is no point in seeking to eradicate racial discrimination in voting. Rather, I believe the point is that, as in all matters of constitutional reform, the search is not for the "perfect" but for the "better" and that, at some point in the evolution of the law, the definition of "rights" requires hard choices, for the exercise of a right by one person may often be perceived as the diminution of the right of another.

Observations

The entire populace of South Africa, not only its citizens of color, will take great strides forward should South African law come to incorporate the principle that oppressive, stigmatic racial discrimination is impermissible and illegal. A key element of such a legal reform will be extending the right to vote to all citizens without discrimination on the basis of race.

But it would be foolish to ignore the plain historical fact that adoption of such legal principles will be the beginning, rather than the end, of defining the contours of race relations law in South Africa. The reasons for the continuing debate over these contours are not peculiar to South Africa, although they are very powerfully experienced there. Wherever people continue to think or act in racial terms, the issue of the permissible scope of private discrimination will remain. Wherever benefits and burdens are not spread equally in society, without regard to race, laws will affect different races differently no matter how sensitive or humane the legislator. And wherever present disparities are obviously traceable, at least in part, to previous overt discrimination by the state, a case can be made for the legitimacy of race-based remedial action.

In deciding to adopt basic anti–race discrimination principles to govern their public and political life, South Africans will enrich themselves and their country and end their international political isolation. In confronting the subsequent issues concerning the precise parameters of these principles, South Africans will stare into the very soul of their nation and will confront, again and again, the questions of just how much they are divided along racial lines and just how far they are willing to go to try to erase those lines.

To some, the quest for a neat set of rules implementing the principle of no racial discrimination is a fool's venture. Certainly, U.S. law has produced no tidy answer to such questions as how to distinguish between permissible private discrimination and that which constitutional or statutory law forbids or how to distinguish between unintentional and purposeful discrimination. South Africa, I have suggested, will find these questions even harder to resolve. Perhaps, then, a reformed South Africa will give up that search for legal controls on racial discrimination and simply continue squabbling—often along racial lines, but with different racial groups holding the power—over the distribution of South Africa's resources. Should this come to pass, it will signal the abandonment of hope for a stable nonracial democracy in South Africa.

Notes

I owe special thanks to Bill Gould, Mike Seidman, and Gerry Spann for very helpful comments. This chapter was inspired by my colleagues at the University of Natal, particularly George Devenish, Karthy Govender, David McQuoid-Mason, and Vincent Mntambo, all of whom are working to instill the reality of justice in South African law. An earlier draft of the first part of this chapter was published in the June 1992 edition of the South African Attorneys' journal, *De Rebus*. This chapter is also scheduled for publication in volume 10 of the *Harvard Black Letter Journal.*

1. The African National Congress has, of course, long advocated such a policy. See, for example, ANC Constitutional Committee, *A Bill of Rights for a New South Africa,* Article 3 (Political Rights), 1990. The National Party has also proposed a voting system in which blacks would be franchised, but additional provisions would dilute the voting power of the majority. See ANC, "National Party Plans for a Future South Africa" (undated mimeo, copy on file with author).

2. In the United States, racial discrimination in voting was outlawed by the Fifteenth Amendment to the Constitution, ratified in 1870. Nine years later, *Strauder v. West Virginia,* 100 U.S. 303 (1879) forbade most racial classifications (except for so-called separate but equal provisions). These events, obviously, did not usher in a democratic, nonracial United States.

3. Thus, this chapter deals with the content, rather than the source(s), of the emerging South African law of race relations. Whether this discussion proves to be about administrative law or constitutional law or legislative options depends upon the resolution of other questions confronting South African constitution makers not taken up here, such as the distribution of authority to make rules among the legislative, executive, and judicial branches.

4. See, for example, *Moose Lodge No. 107 v. Irvis,* 407 U.S. 163 (1972) (dissenting opinion of Justice Douglas): "The associational rights which our system honors permit all white, all black, all brown and all yellow clubs to be formed. . . . Government may not tell a man or woman who his or her associates must be. The individual can be as selective as he desires." The principle that an open, democratic society may tolerate some forms of private discrimination is not boundless. At some point, the magnitude of the harm caused by such discrimination or the transparency of the government's willful refusal to stop it may require that government intervene. See *Shelley v. Kraemer,* 334 U.S. 1 (1948) (holding unconstitutional state court enforcement of racially restrictive covenants on the resale of private housing).

5. In *Palmer v. Thompson*, 403 U.S. 217 (1971), the Supreme Court held that a city council was permitted to close a municipal swimming pool following a desegregation decree.

6. Of course, I do not mean to say that the presence of competitive markets vitiates the utility of or the need for antidiscrimination laws. I do mean to observe that where to draw the public-private line is a difficult and delicate matter that will be affected, inter alia, by the degree of freedom afforded by private markets, organizations, and institutions.

7. This point is based on my own observations of the economy of South Africa while residing there from May through August 1991. Sadly, very little of the literature on South Africa discusses with care the industrial organization policies of the South African (white) government and the effects these policies have had on freedom and initiative in South Africa. A general overview of the extent to which government controls output, prices, and entry in major areas of commerce can be gleaned from L. Thompson, *A History of South Africa* (New Haven: Yale University Press, 1990).

8. 426 U.S. 229 (1976).

9. Of course, the extent to which this conclusion was "logical," as well as what might constitute a valid, racially neutral test of communicative skills as those skills predict effective police work, may be a tightly contested issue. See note 18.

10. See Ely, "Legislative and Administrative Motivation in Constitutional Law," 79 Yale L.J. 1205 (1970).

11. As the following paragraphs show, other—perhaps less racially intrusive—remedies might have been available.

12. See Perry, "The Disproportionate Impact Theory of Racial Discrimination," 125 U.Pa.L.Rev. 540, 558 (1977).

13. 401 U.S. 424 (1971). It is not my intention, in this short piece, to pretend to provide a primer on U.S. race relations law. However, the *Griggs* doctrine was somewhat qualified in *Wards Cove Packing Co., Inc. v. Atonio*, 490 U.S. 642 (1989), partly because of some of the concerns reflected in *Washington v. Davis*.

14. Somewhat hidden in the concepts of "unintentional discrimination" or "disproportionate impact" or "job validation" is the question of what we might mean by racially just employment standards. To the extent that, particularly because of previous discrimination and segregation, educational levels and cultural norms are generally distinct for black people and white people who reside in the same geopolitical entity, there may be no such thing as a generally valid job performance test that treats black and white people equally. A test that measures verbal skills as they accurately predict workplace performance might need to be worded very differently for black and white applicants. Again, this problem is likely to be particularly acute in South Africa, where no single language is spoken by more than 40 percent of the populace and where people of color and white people have been systematically prohibited from intermingling. In a country where, until very recently, black and white people could not ride the same bus or rest on the same stretch of beach, much less attend school together, it is quite likely that cultural differences divide the races in ways that few people understand.

15. For example, one source calculates that in 1985, whites (who constitute less than 15 percent of the populace) accounted for 67.2 percent of all household expenditures, while blacks (who make up more than 75 percent of South Africans) accounted for 20.7 percent. "South Africa," *Economist,* November 3, 1990, p. 9. For even more extreme disparities in land ownership, see same article, pp. 18–22.

16. In *Regents of the University of California v. Bakke,* 438 U.S. 265 (1978), the Supreme Court invalidated a medical school's admissions plan that ensured the admission of a specified number of students from certain minority groups.

17. In *United States v. Paradise,* 480 U.S. 92 (1987), the Supreme Court upheld a trial judge's order that "for a period of time" the Alabama highway patrol must award at least 50 percent of its promotions to corporal to black troopers if qualified black candidates were available.

18. Interestingly, this may be true only if we regard *Washington v. Davis,* 426 U.S. 229 (1976), discussed in the first part of this chapter, as correctly decided. In a world in which disproportionate impact established discrimination (or was substantial evidence of purposeful discrimination), such a rule could be said to be discriminatory in that it would lower the level of spending on white children, who constitute a small percentage (a "discrete and insular minority") of the population.

19. The principle that remedial measures that do not employ express racial classifications are, where feasible, preferable to overt racial preferences is a centerpiece of the current Supreme Court's affirmative action jurisprudence. See *City of Richmond v. J. A. Croson Co.,* 488 U.S. 469 (1989) (Part IV).

20. The argument for such a position is forcefully set out in Ely, "The Constitutionality of Reverse Racial Discrimination," 41 U.Chi.L.Rev. 723, 735–736 (1974). For a recent case in which the U.S. Supreme Court looked more critically at a race-based remedial program in part because it was enacted for the benefit of black citizens by a majority black city council, see *City of Richmond v. J. A. Croson Co.,* 488 U.S. 469 (1989) (Part III. A).

21. See ANC Constitutional Committee, *A Bill of Rights for a New South Africa* (1990).

22. U.S. Constitution, Fifteenth Amendment, ratified 1870.

23. Particularly important is the Voting Rights Act of 1965. That act is discussed in detail, and upheld against constitutional challenges, in *South Carolina v. Katzenbach,* 383 U.S. 301 (1966).

24. In U.S. legal history, the most famous example concerns the long-running battle between the U.S. Supreme Court and the Texas Democratic Party over the latter's attempts to preclude black voters from exercising any influence in the party. This history is summarized in *Terry v. Adams,* 345 U.S. 461 (1953).

25. The U.S. Supreme Court has recognized that U.S. political parties have such rights. See, for example, *Tashjian v. Republican Party,* 479 U.S. 208 (1986).

26. Of course, intermediate alternatives exist. For example, one could divide the city into five distinct districts and select two people from each district. Or one might select some council members from districts and others "at large." Such devices can mitigate, but do not avoid, the instances of apparent discrimination, discussed below, that might arise.

27. This problem is suggested by the facts in *Mobile v. Bolden,* 446 U.S. 55 (1980). Black citizens argued that an at-large, winner-take-all system for selecting city commissioners discriminated against black voters, who were consistently outvoted by the white majority. The Supreme Court, in a supremely ironic result, could not muster a majority of justices who could agree on a single theory to dispose of the case. In *Rogers v. Lodge,* 458 U.S. 613 (1982), the Court affirmed a trial court's decree invalidating an at-large voting system that submerged the minority of black voters.

28. Under the hypothetical plan, five positions are open each year. With ninety black voters for every ten white voters, five black candidates could each re-

ceive eighteen votes for every ten votes garnered by the top white candidate if an
at-large system were in place.

29. This problem is suggested, in part, by *United Jewish Organizations v. Carey,* 430 U.S. 144 (1977). Seeking to enhance black voting strength, New York drew a district line so that a 30,000-member Hasidic Jewish community, previously located in one district, was then spread across two districts. The Supreme Court found no constitutional violation in these overt racial politics because New York was not motivated by hostility toward whites or Jews.

30. Principle (2) is already the law in apartheid South Africa. "Whites," "Coloureds," and "Indians" each have their own separate parliaments and administer their "own affairs." For the liberation struggle that black South Africans have waged to turn out to be a vehicle for, inter alia, substituting black for white dominance over the so-called coloured and Indian races would be a tragedy indeed.

8

National Reconciliation and a New South African Defence Force

Herbert M. Howe

The new South African government will seek a competent military loyal to the nation's constitution. While the new government will include leading African National Congress members, it will have to rely for its military mostly upon the present South African Defence Force (SADF)—a long-time opponent of the ANC. The ANC's military wing, Umkhonto we Sizwe (MK), is small and lacks conventional military capabilities. Many ANC members will suspect the loyalty of the SADF. Even as late as 1992, elements within the SADF considered the legalized ANC as "the enemy" and were mounting covert operations against it.[1] What problems might the new government encounter when seeking a more representative military that includes some MK members, especially at the senior levels, and the demobilization of some present SADF units or personnel?

A representative military that includes both SADF and MK members could greatly assist national reconciliation. Initially, an agreement on a new defense force is necessary for the ANC to accept an overall negotiated settlement. The symbolism of former combatants uniting could aid reconciliation. Significant inclusion of white personnel could reassure the economically important whites about protection of their rights. By drawing from the two opposing forces, the new military would underline a general "no-losers" policy, would decrease sources of potential opposition through inclusion, and would protect the government's economic and political reforms.

An examination of military structure/composition, control, training, and timing issues indicates that the SADF and the MK will accept a merger and that the merger could begin under the Transitional Executive Council. The merger will probably comprise the smaller MK integrating into the larger and more conventionally skilled SADF, but continuing SADF "dirty trick" operations make the creation of a new force, drawing upon ex-SADF, MK, and other personnel, increasingly likely. While the merger will probably include personnel from the homeland armies and conceivably even from other groups (KwaZulu Police and the Azanian Peoples Liberation Army), I will limit my examination in this chapter primarily to the SADF and MK because of their present and future military capabilities and their affiliation with the two major political actors.

Both the SADF and the MK are undergoing unusual, unprecedented physical, personnel, and psychological changes that reflect their declining status. President de Klerk lessened the power of South Africa's securocrats who reigned supreme during the 1980s. Throughout the 1980s, the SADF played a key administrative and decisionmaking role in P. W. Botha's Total Strategy, which marshaled all the nation's political, economic, and coercive tools to combat opposition. The SADF helped run the National Security Management System (NSMS), the State Security Council (SSC) and the Joint Management Committees of the SSC. De Klerk downgraded the SSC and relegated it to its original status as one of four standing cabinet committees. The cabinet, which previously had simply acceded to SSC actions, now has veto power over SSC decisions.

President de Klerk has cut the SADF budget since 1990 and has dramatically downsized the arms industry. ARMSCOR (Armaments Corporation of South Africa) and its affiliates have laid off over 45,000 employees and canceled at least eleven major weapons projects. De Klerk has cut South Africa's standing force from about 75,000 to about 50,000 by initially halving national service to only one year and then by abolishing it altogether. He removed ex-General Magnus Malan as minister of defense and initially replaced him with Roelf Meyer, who reportedly engineered the early departures of hardliners Lieutenant-General Witkop Badenhorst of Military Intelligence and Major-General Joep Joubert of Special Forces. In late December 1992, de Klerk fired twenty-three high-ranking officers for acting, in unspecified ways, against his reform program. Most of the officers had served in the Department of Military Intelligence (DMI), which observers had linked to an ongoing campaign against the ANC. Finally, under de Klerk the Reconnaissance Commandos, an elite special warfare unit with close ties to DMI, went from a full force of about 6,000 down to about 2,500.

The ANC political leadership has suspended internal recruitment and training. Most of MK is still several thousands of miles away, in Tanzania and Uganda, and foreign supporters have largely stopped weapons shipments to MK. These security downsizings may have aided negotiations by reassuring each side—especially South Africa's whites—of the other's motives. The ANC's August 1990 suspension of its armed struggle "marked a watershed in South Africa's history," notes Pauline Baker. It "showed the security-conscious white electorate that de Klerk was engaged with genuine moderates who were prepared to compromise."[2] While the SADF's downsizing initially appealed to the ANC, the apparent involvement of individual SADF personnel in continued urban violence has weakened political reconciliation and angered all levels of the ANC.

Psychologically, each force accepted heavy political indoctrination about the other as its implacable enemy and yet now, after years of opposition, sees its political master negotiating a political settlement. Political resolution of the Angolan and Namibian conflicts damaged the morale of both units. The 1990 suspension of the MK's struggle disappointed many MK partisans "to whom the symbolism of the armed conflict remains important."[3] The SADF had militarily reigned supreme in Namibia (and usually in Angola), but the political settlement forced it to give way to its enemy and forced the MK to leave its only southern African camps for Tanzania and Uganda. Inquiries into the past conduct of both forces have probably hurt internal morale by revealing serious misuses of authority.

Relations between the SADF and the MK have improved during the reformist de Klerk era. By mid-1993 high-ranking personnel from both forces were meeting to discuss their inevitable integration. Continued political compromises between government and ANC negotiators as well as increased personal contact between SADF and MK personnel have significantly lowered mutual suspicions.

Will the government and the ANC agree to a military merger? A laughable proposition during the P. W. Botha era, merger now is a virtual certainty. Top government and ANC spokespersons agree that inclusion of MK personnel will give increased legitimacy to both the SADF and the new government, while the nation will need the competency of existing SADF personnel.

Will the SADF accept a government-mandated merger? The SADF's top echelon enjoyed a generally reformist reputation during P. W. Botha's era and more recently has publicly supported de Klerk's changes and employed physical force, notably twice at Ventersdorp, against white conservatives. It has assented, however grudgingly, to de Klerk's limited restructuring of the defense force. In 1992, Kat Liebenberg, chief of the SADF, acknowledged future MK inclusion in his proposed "Super Defence Force."[4] Therefore, some observers assume an automatic SADF acceptance.

Yet, as Samuel Huntington notes, political reform may transform military beliefs from reformist to conservative: "As society changes, so does the role of the military. In the world of oligarchy, the soldier is a radical; in the middle-class world he is a participant and arbiter; as the mass society looms on the horizon he becomes the conservative guardian of the existing order."[5]

Could South Africa resemble Algeria of the late 1950s when French officers initially obeyed de Gaulle but then, as de Gaulle's policies increasingly forced the officers to act against fellow white conservatives, began to militarily oppose their civilian president? Could the Ventersdorps not prove an affirmation of military loyalty but emerge as a widening crack in civil-military relations? The removal of Generals Malan, Badenhorst, and Joubert as well as the December 1992 firing of twenty-four top officers may have raised civil-military tension. As Jacklyn Cock notes in Chapter 9, the SADF is divided into constitutionalist and praetorian wings, with the latter containing operational army units. Early 1992 polls estimated Conservative Party support among white SADF members at around 60 percent.[6]

Clearly, a widening division between de Klerk and his officers or between Mandela and the MK would threaten both a successful military merger and overall negotiations. Yet several factors will temper overt military opposition to a restructured SADF that includes MK. Despite growing international displeasure about continuing covert operations, de Klerk has moved cautiously on weakening the military's corporate status. Perhaps fearing a variety of hostile reactions from his military or else because he needs a loyal security force if negotiations fail, de Klerk has cautiously restructured his security forces. André du Toit writes that "the securocrats have been willing to go along with the new political initiatives launched by de Klerk, but on the implicit understanding that they will remain masters in their own house."[7] Then, until the December purge, he had eased his personnel changes by seeking reassignments and not resignations and by offering attractive pensions to early retirees. Despite growing evidence of rogue, or praetorian activity from the SADF, de Klerk waited until December 1992 before restructuring his military. He balanced his firing of top officers by apparently alloting them full pension benefits and by not pressing any legal charges against them.

While some SADF officers regard the force as a "guardian of the volk" (and the SADF officer corps is about 80 percent Afrikaner), the strong March 1992 referendum vote for de Klerk's reforms dampened military thoughts of opposition. Over 60 percent of the Afrikaners (and 68 percent of all whites) supported de Klerk. De Klerk's popularity has fallen significantly since then, but no opposing individual or organization has gained wide acceptance as an alternative to de Klerk and his reforms. Interviewed SADF officers acknowledge a war-weariness and the inevitability of a political change. Continuation of the status quo would prolong the nation's negative growth rate and growing violence while reinviting sanctions.

MK will accept merger of its small force into the much larger SADF. MK has generally been regarded as one of the ANC's more radical forces; given its ongoing drop in status and presumed opposition to the ANC's compromises with the government, could MK actively oppose merger with its former foe? Stephen Davis, author of *Apartheid's Rebels,* notes that MK has publicly supported the ANC's attempts at peaceful negotiation and even the Pretoria Minute, which suspended armed violence. "There's no real evidence of a split on the negotiations issue," says Davis, "only over the specifics."[8] Marina Ottaway suggests that the ANC and MK favor inclusion: "Disbanding MK without obtaining its integration in the SADF would involve new risks [from township youth]."[9] By mid-1993, the ANC was losing control of the returned MK cadres, who were increasingly protesting the lack of jobs and turning to violent crime. As with possible rogue SADF units, the only threat from rebellious MK personnel would be that of provoking limited hostilities that would slow down negotiations.

Structure and Composition

Will MK form a new army along with SADF personnel or will it merge with the present SADF? What proportions of SADF versus MK personnel might the reconstructed military include? What might the size of the new army be? Should the new government conduct loyalty checks, based on past behavior, or should it adopt a general amnesty policy? Two realities— one military and one political—will largely determine the structure/composition of South Africa's new military.

The SADF is a far superior military force. The SADF outnumbers MK, having a standing force of about 45,000 (Permanent Force and National Servicemen) and an estimated Citizen Force and Commando reserve of about 360,000. Recent estimates place the present MK membership between 4,000 and 9,000. The SADF is a modern mechanized army well trained and experienced in both conventional and counterinsurgency warfare. Until recently, very few MK officers had received any significant conventional training, and few MK have seen military combat. Leading ANC officials, including the late Chris Hani, former head of MK, have acknowledged the professional competence of the SADF and that a new South Africa would be wise to retain much of the existing SADF.

The liberation struggle for South Africa was largely a political struggle. Unlike Mozambique, Angola, Zimbabwe, and Namibia, an insurgent military played a relatively minor role. Given its military weakness and its lack of results, MK has less of a bargaining position, both within the ANC and with the government, than did ZANLA with ZANU in Zimbabwe or FALA with UNITA in Angola.

Partially balancing the SADF's military capabilities will be South Africa's new political reality of black majority rule. The composition (and conduct) of the new security forces must more accurately reflect South African society as a whole. The SADF has publicly favored only integrating limited numbers of MK personnel into existing structures. Yet integration of a limited MK into the largely unchanged SADF would present overtones of continuing white hegemony. MK supporters would enter existing units and receive the uniforms, emblems, and mottoes that symbolized the previous white supremacy. The ex-MK members, given their general lack of conventional training, would be commanded by existing SADF officers. Such a situation of "the more things change . . . " would decrease the new government's legitimacy to many blacks.

Some ANC officials have suggested demobilizing the present SADF and creating a new military that would select and then train ex-SADF and MK personnel. Yet South Africa's continuously pressing security needs (in mid-1993, 10,000 troops were patrolling the townships) precludes forming a new army. Most likely, a future SADF will retain its existing structures but provide nonoffensive symbols (e.g., new unit names, uniforms, emblems).

Whether South Africa's future military is an integrated or a new force, present MK members will initially play a more symbolic than substantive role. Both the SADF and MK are presently jockeying for maximum participation. The conventional and well-trained SADF is increasingly nonwhite: by early 1992 it was about 35 percent nonwhite compared to about 10 percent in the early 1980s. Cock, in Chapter 9, states that the Permanent Force now is 52 percent nonwhite. Recent volunteer intakes included only a minority of whites (and less than a third of eligible whites answered their conscription call-up). In August 1993 the government announced that national conscription, which had included only whites, would end. By the time an interim government begins the process of merger, the active-duty SADF could be at full strength for peacetime with an overall trained force of about 50 percent nonwhite, with no formal allegiances to the ANC. From the approximately 10,000 members of the SADF-trained homeland forces should come additional non-MK enlistees.

To gain public acceptance, the new military will have to appoint some MK officers into high positions and offer enlistment to former MK supporters. MK can reasonably argue that to help legitimate the new government, the military must include a reasonable number of ex-MK, and offer an affirmative action policy and accelerated training programs. The new South Africa will face no serious external threats, and the political benefits of ex-MK participation will outweigh any temporary losses of efficiency.

To meet the requirements of South Africa's new military (and to help ensure its own survival), MK has stopped its guerrilla training and is seeking conventional officer training from foreign sources. Reportedly, Tanzania now offers only conventional training to some 3,500 MK members. Other countries apparently training MK, in small numbers, include India (naval officers) and, at least until recently, Ethiopia (pilots).

Given MK's small size, its initial representation will appear more noticeable in a few high-ranking positions than within the enlisted ranks. Importantly, the ANC appears willing to accept the SADF (and homeland units) initially dominating the new military. The MK has established a welfare program for ex-members to aid their adjusting to civilian life and to provide them with job skills.

The former SADF will initially control the air force and navy (the ANC has received minimal training in these two branches) and, along with the homeland armies, will staff most of the army's specialist grades as well as the majority of the army command and control positions.

As with other Southern African states, South Africa will likely have a relatively small peacetime army. South Africa's major security problems, resulting from the legacy of apartheid coupled with rising expectations, arise from within South Africa's borders. The South African police force will continue to grow—some experts believe that by the year 2000 the force will be over 130,000 (versus about 35,000 ten years ago)—and thus can accommodate potential enlistees. The civilian economy, were it to revive, should begin absorbing potential enlistees and thereby lessen pressure to convert the military into an employment agency. The new government will face extraordinary "peace dividend" pressures and thus continue the present downsizing of its defense force. The ANC prefers a small army also because it would lessen the numerical disparity between the present MK and SADF.

While reserve units will be less necessary, the new government could retain, at least for an interim period, a restructured Citizen Force. This nonracial system could absorb unemployed soldiers, offer education to above-age secondary school students, engage in civic action programs, and function as a local militia force backing up police units.

As for loyalty, both the SADF and MK publicly state that the new government should not use past political loyalties (or actions) to bar individuals from the new defense force. The ANC has indicated a willingness for a future amnesty. Yet several SADF units will either disappear or face overhaul. Special Forces units are most at risk, because of both their past actions and their future threat during the transition. *Africa Confidential* claims that covert units like the Civil Co-operation Bureau (CCB) "have their own structures and, often, sources of funding and are composed of individuals for whom violence is a way of life. Quite literally, they have a vested interest in destabilization. It is their job."[10] Annette Seegers believes that soldiers in covert units have a "'special operations mentality,' here meaning that some soldiers could view themselves as somehow above normal military ethics . . . that only extreme measures could prevent political demise."[11] Several observers suggest that special units outside of the normally tight chain of command have profited from sales of ivory, diamonds, and hardwoods.[12]

Having borne the brunt of destabilization's battles, these Special Forces units might see their ideological world (Afrikaner dominance and ANC dislike), their military identity (covert/ Special Forces groups), and possible personal aggrandizement threatened by a changing political universe. De Klerk has disbanded the 32 Battalion and Koevoet (although its members still remain with the security forces). Other SADF groups at future risk include the presently structured Reconnaissance Commandos and Military Intelligence, both of which de Klerk has downsized through attritions or firings. Numerous observers, as Cock notes, link these groups to the perceived "third force," whose violent actions have threatened negotiations. Military Intelligence still retains access to the publicly unaudited Special Defence Fund of $1.5 billion. Presumably the longer these groups aid domestic political violence, the less chance of their survival under a new government.

Control

For the new military to serve as a force for national reconciliation, it must impartially adhere to constitutional principles. Yet neither the SADF nor the ANC has fully risen above partisan politics. Opponents of apartheid have sometimes identified the SADF as the enforcer of the National Party's regional destabilization and as a policer of the townships in the late 1980s. Critics of the MK note the use of political commissars and wonder whether the ANC will find difficult the transition from obeying (and directly influencing) a political movement to unquestioningly enforcing statutes. Both sides received prolonged political indoctrination that cannot be quickly wished away.

The dichotomy of a largely black-elected government and a white-staffed military curiously may lessen potential political interference. Both sides will gain from a nonpolitical force. The whites understandably will seek a professional, apolitical military, while the government equally will avoid provoking a force initially dominated by officers once trained to see the ANC as the nation's enemy.

A new government might not have to worry about SADF loyalty, especially if the interim government demobilizes the Special Forces units. The SADF traditionally has entered politics only when requested. As Jakkie Cilliers notes, "The SADF is essentially a professional, bureaucratic-type institution with strong internal control."[13]

In February 1991, de Klerk revoked the decree exempting SADF officers from trial for murders conducted in operational areas. In October 1992, both sides agreed that the new SADF should be above politics when they agreed to a draft code of conduct as part of the National Peace Accord. Inter alia, the code holds that individual soldiers are legally responsible for their actions. They may thus disobey orders that clearly are unconstitutional. In the short run, despite the serious lack of enforcement mechanisms, as noted by Cock, it is hoped that the code indicates a growing acceptance by SADF planners of the need for a constitution-obeying military.

Training

Military training not only molds previous conventional and guerrilla forces into a new conventional force but should also instill loyalty to the constitution rather than to a political movement. In Southern African states—notably Zimbabwe (1980), Namibia (1990), and Angola (1992)—where all armies remained standing at war's end, the government called in politically neutral foreign trainers.

South Africa's situation has few, if any precedents. Political and economic pressure forced the government to enter negotiations, although its security forces effectively prevented MK from posing any serious threat. The SADF feels fully competent to train whatever MK or Inkatha personnel it might integrate. Yet non-SADF training of a new military would lessen MK's sensitivity about being absorbed by the erstwhile enemy. The ANC has approached other countries, e.g., Britain, France, and India, as future and politically acceptable trainers of the new military. The most likely external candidates are a Commonwealth force or Britain, whose British Military Advisory Training Team (BMATT) has an extensive training record in Southern Africa. Such involvement would initially focus upon infantry basic training. Elements of the homeland armies could serve a supplementary training role. Racial factors probably increase their acceptability to MK and its followers, and their SADF training and their non-British citizenship could make them acceptable to the largely Afrikaner officer corps.

Timing

The existing South African government will retain full military control until the Transitional Executive Council (TEC) begins, probably by late 1993. The TEC will exercise oversight of the SADF but will leave operational responsibilities to the defense force. Between late 1993 and April 1994 the TEC, acting along with the South African government, will coordinate plans for integration of MK with the SADF. A likely precursor to full integration could be a National Peace Keeping Force, comprising the SADF, MK, and the South African police. After the 1994 elections, the new goverment will assume full control over the SADF.

Unforeseen difficulties could still undermine negotiations and forestall a military merger. More likely, however, both sides will accept a merger (an integrated force is more likely than a new one) and contribute to national reconciliation. The force will be smaller and less funded than before and certainly less involved in regional conflicts. The better-trained SADF and homeland army members will initially dominate the new military, with only limited representation by MK and even less by Inkatha and other groups. The new government should have little reason to disregard the traditional division between civil and military relations. Demographic differences, accelerated training, and early retirements should see blacks, both ANC and non-ANC, dominate the lower and then the middle ranks of the new SADF.

Notes

1. "SADF Plan to Counter ANC 'Enemy,'" *Weekly Mail,* May 15–21, 1992.
2. Pauline Baker, "A Turbulent Transition," *Journal of Democracy* 1, no. 4 (Fall 1990), p. 12.
3. Marina Ottaway, "The ANC in Transition: From Symbol to Political Party," *CSIS Africa Notes,* no. 113, June 20, 1990, p. 5.
4. Ibid.
5. Samuel P. Huntington, *Political Order in Changing Societies* (New Haven: Yale University Press, 1968), p. 363.
6. "South Africa: Violent Reactions," *Africa Confidential,* February 21, 1992.
7. André du Toit, "The White Body Politic: What Has de Klerk Wrought?" *The Southern Africa Policy Forum Report* (Queenstown, Md.: Aspen Institute, 1991), p. 30.
8. Interview with Davis, April 15, 1991.
9. Ottaway, "The ANC in Transition."
10. "The Rule Of Law," *Africa Confidential,* May 3, 1991.
11. Annette Seegers, "After Angola and with de Klerk: Current Trends in South Africa's Security Establishment," in Vernon J. Kronenberg et al., eds., *The Military and Democracy* (forthcoming).

12. "SADF Hits at Press over Ivory," *Cape Times*, May 26, 1989.

13. Remarks prepared for the Institute for Democratic Alternatives in South Africa conference on a future SADF, May 23–27, 1990.

9

The Dynamics of Transforming South Africa's Defense Forces

Jacklyn Cock

A ny discussion of the dynamics of transforming South Africa's defense forces must be anchored in the political context within which the SADF and MK operated until recently.[1] South Africa is only just emerging from the period of "total onslaught." This ideology was the launch pad for the militarization of the entire society as the state mobilized resources for war on political, ideological, and economic levels.[2] This process was spearheaded by the SADF.

The SADF has always been used to maintain the apartheid regime and white minority rule. It has done so inside the country with a policy of violent repression and has engaged in an undeclared war of destabilization against neighboring states that has been described as a "holocaust."[3]

The process of creating a legitimate and representative defense force reflects all the political difficulties involved in the transition to a nonracial democratic order in South Africa. Current attempts to understand and debate the process are flawed by a number of analytical problems.

Analytical Problems

The current debate on the formation of a new defense force in South Africa has focused narrowly on the integration of the existing armies—the SADF, MK, and the homeland forces—into a single defense force. This conceptualization neglects two more crucial aspects of integration. First, soldiers must be incorporated into the wider society. This applies both to

MK cadres returning to South Africa and SADF soldiers who are being re-trenched or who prefer not to serve in a future nonracial defense force. Second, these two aspects of integration—of armies and of soldiers—should be related to the larger issue of the integration and transformation of the entire society. Our central task in South Africa is to build a common society and create institutions that unify rather than divide us.

The next analytical problem involves contextualization, in both the political and historical senses. The question of a new defense force must be firmly located in the context of the current negotiations on a new con-stitutional dispensation.[4] The debate must also be located within the process of restructuring the SADF and the arms industry, which has al-ready begun. Since 1990, mandatory military service has been elimi-nated, defense spending has been cut, programs and personnel of the SADF have been reduced, various SADF bases have been closed, the Na-tional Security Management System has been dismantled, ARMSCOR has been privatized, and several ARMSCOR projects have been can-celed, leading to the retrenchments of almost 45,000 workers in the arms industry.[5]

This points to a further analytical problem in the debate about the new de-fense force. The current focus on the possibility of a coup d'état by the SADF in the *future* diverts attention away from the covert, clandestine ways in which the SADF is already intervening in the political process *in the present*.

Another analytical problem is the tendency of analysts to assume that the ANC and the government are monolithic entities. Both MK and the SADF are extremely fractured institutions. MK is fractured geographically with its members dispersed into three categories: those inside South Africa in underground structures (often involved in defense units);[6] about 3,000 in camps in Uganda and Tanzania; and a third category undergoing train-ing in India, Cuba, and other overseas countries.

The SADF is also a fractured institution but along ideological lines. While there is no systematic sociological evidence, there are reports that the senior officer corps of the SADF is deeply divided between "constitu-tionalist" and "praetorian" factions, the latter clustered around the Depart-ment of Military Intelligence/Special Forces axis.

There are also ideological differences within MK. Some of these have surfaced in relation to future defense policy. There is a militarist tendency that is eager to maintain a powerful defense capability and not to erode the technological capacity South Africa's arms manufacturers have devel-oped.[7] There is also an antimilitarist tendency that is illustrated by the statement of John Nkadimeng, who outlined some basic principles of MK's future defense policy at the Lusaka conference in 1990 on the future of security and defense in South Africa, jointly hosted by the ANC and the Institute for a Democratic Alternative in South Africa (IDASA).[8] "The ultimate objective of our society should not be to build more barracks but more schools and hospitals. It should not be to manufacture more AK and R1 rifles but more tennis rackets and golf clubs. Not more tanks and Hippos [military vehicles] but more tractors and harvesters."[9]

There are also differences within MK on the appropriate model of military organization for a future defense force, with some officers attracted to a Western professional model with its emphasis on the corporate neutrality of the armed forces and many cadres attracted to a more populist ideal along the lines of the professional revolutionary model of civil-military relations as in China or Cuba. There is a good deal of romantic populism among MK cadres.[10]

This kind of differentiated analysis is also necessary to offset the tendency to portray the SADF as a racially homogeneous institution. In reality, 55 percent of the 43,000 members of the SADF Permanent Force are African, coloured, and Indian. However, it remains a white-dominated and white-controlled organization. The most senior black officers in the SADF are three colonels (compared to ten women), and all brigadiers and generals are white.

A further flaw in debates about the new defense force is that reliance on public utterances blocks developing a more nuanced analysis that takes account of different factions and competing positions within the two armies. Furthermore, the responses of the various actors are assumed to be static. In fact, they have changed significantly in recent times.

A final difficulty in analyzing the dynamics of transformation to a new defense force concerns access to information. The SADF has always operated in a sealed environment shielded from public knowledge and scrutiny. A mass of laws and emergency regulations hides its powers and activities behind a veil of secrecy. By necessity, as an underground structure, the operations of MK are also shrouded in secrecy.

Despite these analytical difficulties, there are four reasons to be optimistic about the possibility of creating a new, legitimate, and representative defense force.

Grounds for Optimism

First, the present SADF is itself the product of a process of integration of armies previously at war, a process achieved in 1912 without international supervision. "The early Union Defence Force (UDF) represented a synthesis of diverse military traditions and structures—a conventional armed force on the one hand and a guerrilla type army on the other."[11] At the time, it was agreed that in the interests of white unity the integration should involve equal numbers from the opposing British and Boer forces in both the rank and file and the command structures. This would appear to be a powerful model from our own history that we could draw on.

The second reason for optimism is that the most vociferous opponent of integration was demoted and has now retired from politics. The previous minister of defense, General Magnus Malan, emphatically rejected the possibility of integrating the SADF and MK on numerous occasions. For example, in May 1990, he stated: "The SADF is an instrument of the state which protects the security, life and property of all people. On the other hand, MK acknowledges that it is a revolutionary organization. It conducts the revolutionary struggle against the population and aims to destroy that part of society it disagrees with."[12] Malan was emphatic that MK members would not fit into a professional defense force because of different forms of training and the technologically advanced level of the SADF.[13]

A third reason for optimism about the creation of a new legitimate and representative defense force concerns the experience of two of our neighbors. The transitions to independence in Namibia and Zimbabwe involved the merging of government and guerrilla armies in the establishment of new national defense forces.[14]

Another factor providing grounds for optimism is that there seems to be quite a considerable level of agreement among different political actors that the new defense force should be a constitutional army working within the rule of law; that it should be nonpartisan and nonracial; that its primary role should be to protect the territorial integrity of South Africa; and that it should be subordinate to civilian authority and fully accountable to parliament. At the MK conference in August 1991, the seven hundred delegates declared in favor of a professional nonpolitical army answerable only to the constitution.[15] There are officers within MK, the homeland armies, and the SADF who support the notion of a Western "professional" model with its emphasis on the corporate neutrality of the armed forces.

A surprising level of agreement emerged at the 1990 Lusaka conference. The joint proposals that emerged from this conference of senior ANC leaders, homeland army officers, and an unofficial SADF delegation included the following:

1. There should be a mutually binding cease-fire prior to the commencement of negotiations.
2. The future defense force should be nonracial, open to all citizens.
3. The new defense force will be a professional-type organization with high standards of efficiency.
4. In anticipation of integration the SADF and MK should initiate "politically sensitizing programs."
5. A joint commission made up of defense experts and SADF and MK leaders should be formed to investigate the process of integration.

6. An all-party agency should be established to monitor the security forces in the transition period.

7. Once negotiations have commenced, the return of ANC soldiers to South Africa should be organized in a formal manner.

8. The system of conscription should be phased out and replaced by a professional permanent force and a volunteer reserve.

9. There should be a "gradual but substantial" reduction in force levels of up to 50 percent.

Overall the atmosphere was one of reconciliation and commitment to the future democratic South Africa. (Incidentally, the conference also illustrated the unifying power of patriarchal ideology and practices. Outside the formal sessions there was much socializing among former antagonists, and this took place along conventional patriarchal lines of heavy drinking, discussions of favorite weapons, early morning jogs, and so on.)

Obstacles to Transformation

The Political Intervention of the SADF

The major obstacle to the transformation to a new, legitimate, and representative defense force is the political intervention of the SADF. This intervention ranges from covert and clandestine activities to disrupt the present negotiation process to the possibilities of a coup to overthrow de Klerk, or noncooperation with a new civil authority.

There are reports that de Klerk's position is threatened by a cabal of high-powered generals in the SADF who are implacably hostile to the ANC and whose resistance to reform is growing. F. W. de Klerk's purge of twenty-three officers from the SADF did not include General van der Westhuizen, who remains head of Military Intelligence despite evidence suggesting he is linked to political assassinations. According to John Carlin of the *Independent*, talk of a coup is widespread in Pretoria among police and army officers as well as civil servants. However, Mike Hough of the Institute of Strategic Studies and Helmut Romer Heitman, *Jane's Defence Weekly* correspondent, maintain that top generals are more concerned with issues of financial security and pending retirement packages. The issue of job security within the military is seen by Hough as a potential breaking point: "The potential of dissatisfaction turning into militancy would increase if there were a direct threat to job security through further dramatic cuts to the military budget or through racial integration in the army."[16]

It is known that right-wing sentiments supporting the Conservative Party and the Afrikaner Weerstandsbeweging (AWB) are strong within sizable sectors of the SADF's Citizen Force and Commando units. There is some evidence of links between this level of the SADF and the growing militant white right wing. A total of thirteen paramilitary groups in the right wing have been identified.[17] According to the AWB leader, Eugene Terreblanche, many AWB commando members are SADF trained.[18] The numbers involved are small, but a tightly organized group of well-trained people with access to weaponry could cause considerable disruption and political instability.

It seems certain that there is a very reactionary element within the SADF and the South African Police (SAP) that is intent on destabilization, the weakening of the ANC, and the disruption of the current negotiations. Since 1990, the ANC has frequently claimed that these security elements, referred to as a "third force," were engaged in a covert, clandestine campaign that involved committing acts of terror, training and arming Inkatha members, and fueling township violence. There is mounting evidence that this third force is located in the Department of Military Intelligence, which is responsible for the operations of SADF Special Forces. A number of former agents have come forward to confirm these allegations. For example, Major Nico Basson, a former military intelligence officer in the SADF, has claimed that the SADF was arming Inkatha members and that the assassination of ANC activists—sixty assassinations in 1991 alone— and the random killing of black train commuters—112 killed in attacks on trains on the Reef during the eighteen months ending on January 31, 1992—were part of an elaborate government plan of disruption.[19] Since that time, a number of individuals have given evidence to the Goldstone Commission on extensive linkages between the SADF, Inkatha, and township violence.

In January 1992, then-Defence Minister Roelf Meyer said that neither he nor President de Klerk knew of a third force operating to discredit the negotiation process. "But there are those elements who would like to disrupt the process."[20] Malan himself admitted that he was concerned about the loyalty of some Commando and Citizen Force units, but believed 99 percent of his top officer corps were loyal to the government's reform process.[21]

In the SADF there is a tradition of covert intervention rather than independent action. In a perceptive piece in the *Weekly Mail*, Anton Harber dismisses the threat of a coup as "paranoia." The threat takes insufficient account "of the way key elements of the SADF, in particular DMI and the Special Forces, have always operated: not seizing direct control but using their powers, covertly or overtly, to manipulate and control the political situation. In Zimbabwe, Namibia, Mozambique and Angola, they developed surrogate forces and used a sophisticated combination of training, provision of arms and propaganda to weaken the civilian powers." Harber points out that this is precisely the strategy that is being followed in South Africa now. Coup speculation detracts attention from this process.[22]

A Leadership Crisis

One of the most serious obstacles to the creation of a new, legitimate, and representative defense force in South Africa is the existence of a leadership crisis within the two institutions central to the formation of this new army—the SADF and MK.

There are reports that since General Malan's departure the SADF has been in a political vacuum characterized by an acute lack of confidence in the future and an overall lack of any clear political direction. Brigadier Ramushwana, head of the Venda Military Council, warned in 1991, "Military personnel cannot be left in a vacuum. . . . By adopting a wait and see attitude South Africans run the risk of pushing the existing military forces in South Africa into the role of typical interventionist forces."[23] This lack of political leadership within the SADF's General Staff is especially serious at the present time because direction is needed to bring the SADF within the terms of the current negotiations. At the same time, there has been an equally serious development within MK—first, the resignation and then assassination of the MK chief of staff, Chris Hani. A thoughtful man (his favorite writer was Jane Austen), Hani's public statements in the twelve months before his death displayed a clear commitment to compromise and reconciliation. These statements also demonstrated an acceptance that the SADF demand for the maintenance of "professional standards" must be met. Hani was enormously popular among MK cadres. His resignation created a "crisis of leadership" within MK, and his assassination will further alienate many MK cadres who are dissatisfied and feel marginalized from the current negotiation process.

Differences in Skills and Force Levels Between the SADF and MK

Herbert Howe in Chapter 8 points to the significant differences in size, skill levels, capability, and technical sophistication among the SADF, MK, and the homeland armies. The SADF is the most powerful force in Africa, with the capacity to mobilize a force of almost 500,000, extensive battle experience, and technically advanced weaponry and equipment. The MK by contrast is a comparatively small and ill-equipped guerrilla-type army.

However, there are reports that the ANC is attempting to address these disparities. In October 1990, Chris Hani announced that hundreds of MK recruits had left South Africa to receive training in regular warfare in several countries abroad. Again in July 1991, he was reported as saying that

MK was in the process of being transformed into a conventional army: "Only a professional army will be competent to man a future democratic order. MK is preparing to be part of this."[24]

Tokyo Sexwale, MK commander and ANC head of special projects, has emphasized the urgency and importance of the "upgrading and recruiting of MK soldiers" in preparation for this professional army. "We have 15 different armies running around South Africa. If we don't integrate them soon there will be carnage here—a bloody civil war."[25] Another MK official, Calvin Khan, stated: "We are training a regular army in order to participate fully in the new army of a future South Africa—both in terms of professional ability and to ensure that the new army will not be swamped by an all-white officer corps."[26]

This "swamping" is a very real possibility. It is uncertain whether overseas training will substantially alleviate the problem of disparities. Laurie Nathan has argued that the inequalities in size and capacity between the two armies are likely to result in the absorption of MK into the new defense force that will continue to be dominated by white SADF officers.[27] On the other hand, Rocky Williams has suggested that the differences are not as absolute as Nathan suggests. He points out that part of the "total strategy" project was the extensive recruitment of African, coloured, and Indian soldiers and seamen in addition to the creation and arming of the homeland defense forces (with present force levels of about 10,000 personnel). He refers to a current internal SADF microstrategy "to create an ideological and political counter-balance within its ranks to the eventual incorporation of MK and rebellious homeland 'defense force' members into its ranks. This strategy is evident in the SADF's 'Africanization' of the lower rungs of the SADF and the rapid promotion of African, 'Coloured' and Indian soldiers into positions of responsibility at these levels."[28]

The Highly Politicized, Partisan Nature of Both Armies

A major stumbling block in the creation of a new defense force is the state's conception of MK as a private army that should be immediately disbanded instead of simply suspending its activities. At a press briefing in 1992, Roelf Meyer said, "An organization which was still engaged in an armed struggle was prohibited from a military viewpoint, from becoming part of the SADF."[29]

To many South Africans MK is not a private army but a people's army, and the SADF is the private army or military wing of the National Party. Williams has cited examples of the close relationship between the SADF and the National Party.[30] As Nathan writes, "Over the past two decades the military played a coercive internal and external role, aligned to a particular political party, hostile to others, with a distinct ideological agenda and with the explicit aim of preventing the realization of a nonracial democracy."[31] There is a sense in which both armies were partisan. Even an editorial in the *Sunday Star*—a newspaper not renowned for its radicalism—stated in an editorial in 1990: "It is plain to see why the SADF and Umkhonto must eventually merge. Just as with Umkhonto, the SADF has become a partisan army in the eyes of the majority in this country. The army is seen as actively assisting in the enforcement of the Government's discriminatory laws."[32] Hani admitted that both the SADF and MK were partisan:

> Yes, MK is an army of the ANC and therefore it's in a way partisan in the same way as the SADF is partisan. The new army must contain the best of MK and the SADF. We don't dismiss the SADF and any other armies that might be around. But it must be clearly defined so that the new army won't be used by any political group to entrench a political party in power. The army must also be non-partisan and non-political because when we say we believe in pluralism and a multiparty government, this means the army must not have a problem about serving whatever party is voted into power.[33]

Stages in the Transformation

Given these obstacles, it is difficult to write with any certainty about the pace and trajectory of the formation of a new defense force. However, it is clear that the crucial questions concern the control of the military in the current fluid and uncertain period of transition, and the subordination of the military to the civilian control of a future democratically elected government.

There is a strong argument that the seeds of a future defense force must be sown during the present transition process. Orientation programs and confidence-building measures should be immediately instituted for all soldiers. But in the immediate term, while multiparty negotiations continue, the most urgent requirement is the control of the security forces.

At the time of this writing there is no agreement on a code of conduct for the SADF. The code proposed to the National Peace Committee involved greater civilian control. The cornerstone of the proposals is a defense council to be appointed by the state president made up of experts from across the political spectrum who are acceptable to all parties in the negotiating process. In addition, the proposals call for the appointment of a military ombudsman to investigate irregularities and human rights violations within SADF ranks.

Continuing SADF resistance to these proposals raises the question of whether even a multiparty interim government can manage the security forces. An increasing number of people believe that there will have to be some level of assistance from the international community in managing the process of transformation.

The international community could also play a valuable role in helping to establish some kind of regional security arrangement.[34] Discussions in Africa of such arrangements have stressed the connections between the four "calabashes" of security, stability, development, and cooperation. Current threat analyses focus on the variety of security problems in the Southern African region brought about by the effects of poverty, SADF destabilization activities, drought, disease, and social dislocation. Furthermore, there is an urgent need for socioeconomic reconstruction in South Africa. The new army could play a part in the orderly redirection of resources presently employed in military activities to development ends. Some form of conscription could be considered in a society where 66 percent of the population are under the age of twenty-seven and have experienced an inadequate (either in material or ideological terms) education, and 48 percent of the potentially economically active population are unemployed. Such an arrangement could contribute to nation-building as well as the teaching of functional skills.[35]

Conclusion

The scenario suggested by some, that the future involves a choice between a totally new defense force and the maintenance of the current SADF as the basis of any future defense force, is problematic. The unification of government, homeland, and guerrilla forces is inevitable. The key questions are whether the future will involve a balanced integration of these forces in a way that will contribute significantly to building national unity and reconciliation or, as Nathan and Jakkie Cilliers argue, whether the SADF will maintain its dominance.

Cilliers argues that the future military "will be built around and upon the SADF as it exists today." In his view, the new defense force is likely to be "still run by whites and its military culture will essentially be that of the SADF for many years to come."[36]

Nathan has warned of the "absorption" of MK cadres into the institutional structures of the SADF. In his analysis, continued SADF control of the new defense force could mean the continuance of the military's potential to destabilize and disrupt; a low level of legitimacy in the perceptions of the masses; and feelings of suspicion and insecurity in Southern Africa as a whole, particularly among the Frontline States. Countries like Mozambique and Angola that have had to divert vast resources to defense may be inhibited from substantially reducing force levels and military spending.[37]

The stakes are high for all of us in Southern Africa. As Simon Baynham has pointed out, "In societies deeply divided by race, ethnicity or other primordial affiliations, the composition of the security forces is of vital importance to the state and its inhabitants."[38] The social composition and leadership of a future defense force MUST reflect that of the wider society. Similarly, the ideology of the defense force must reflect those values of the society that the new army is committed to defend. This underlines the difficulties involved in creating a unified defense force in a society that is white-dominated, scarred by deep inequalities, and riven by competing and apparently irreconcilable ideologies. Our main struggle is to create a common society in South Africa. The creation of a united legitimate defense force is one aspect of this task.

Notes

1. My understanding of these issues owes a great deal to discussions in the Military Research Group. I am especially indebted to Laurie Nathan and Rocky Williams.

2. Jacklyn Cock and Laurie Nathan, eds., *Society at War: The Militarisation of South Africa* (New York: St. Martin's Press, 1989).

3. Phyllis Johnson and David Martin, *Apartheid Terrorism* (London: Commonwealth Secretariat, 1989), p. 11. It is important to remember this because some influential analysts refer to the security problems now faced by the Southern African region without any acknowledgment of the role the SADF played in creating those problems in the first place. For example, see J. Cilliers, "Integrating the Military into Democracy: The South African Soldier as a Citizen in Uniform," paper presented at the DISA conference "Southern African Security Relations Towards the year 2000," Pretoria, 1991, p. 3.

4. The issue continually plagues the negotiations process. The status of MK was the subject of a major row between de Klerk and Mandela at the opening session of Codesa. De Klerk argued that an organization that remained committed to armed struggle could not be trusted and demanded that the ANC "terminate" MK before it could enter into binding agreements at Codesa. Mandela replied that to do so in the current climate of political violence would be to commit suicide. The ANC has repeatedly stated that armed struggle (presently suspended) will be abandoned only under an interim government.

5. *Weekly Mail*, October 18, 1992.

6. It was resolved at the ANC 1991 conference that MK should "act in the defence of the people" against so-called third force violence by operating and training self-defense units answerable to local civic structures. These are paramilitary organizations. See J. Cronin, "For the Sake of Our Lives: Guidelines for the Creation of People's Self-defence Units" seminar paper presented at the Project for the Study of Violence, April 1991. There has been no systematic social research on these self-defense units, but B. Xeketwane has reported the fear expressed by 75 percent of the township residents he interviewed in the area of Soweto near Merafe Hostel, that in the long run the defense units prolonged the violence and worked against the creation of a culture of political tolerance. B. Xeketwane, "The

War on the Reef: The Political Violence in the Reef Black Townships Since July 1990." B.A. Honors Diss., University of the Witwatersrand, 1991, p. 72.

7. See, for example, the statement of Tokyo Sexwal in the *Weekly Mail,* October 18, 1991.

8. This conference "The Future of Security and Defence in South Africa," was attended by a range of military actors, including MK and individuals close to the SADF.

9. Remarks prepared for the Institute for Democratic Alternatives in South Africa Conference on a Future SADF, May 23–27, 1990.

10. See H. Barrell, *MK: The ANC's Armed Struggle* (London: Penguin, 1991).

11. R. Williams, "Taming the SADF: How to Build a Legitimate Defence Force," *Mayibuye*, March 1992, p. 32.

12. Malan, as quoted in the *Citizen*, May 28, 1990.

13. "The type of army that is planned for the future is of such a nature that it is doubtful that MK's man who handles an AK (assault rifle) will feel at home there. Weaponry such as the G5, G6 (cannons) and other are exceptionally advanced technically. The difference between the SADF and MK is not just the level of training. The SADF creates technology, MK is simply a user. We are not on the road to using the army to keep unemployment off the streets." *Cape Times*, May 18, 1990, cited in Laurie Nathan, "Riding the Tiger: The Integration of Armed Forces and Post-apartheid Military," *Southern African Perspectives*, Working Paper Series No. 10 (Bellville, South Africa: Centre for Southern African Studies, 1991), p. 3.

14. However, Nathan, ibid., has pointed to important differences between their situation and ours. In South Africa, unlike in Namibia and Zimbabwe, there has been no formal cease-fire, and in both those countries the transition to independence was internationally supervised.

15. *Weekly Mail*, August 16, 1991.

16. Mike Hough and Helmut Romer Heitman, as quoted in the *Star*, February 25, 1992. This implies that the introduction of generous demobilization benefits, including perhaps a GI Bill of Rights–type measure for all veterans—both from the SADF and MK—could be important. (I am grateful to Alison Bernstein for pointing out the significance of this piece of legislation to North American Indians.)

17. *Star*, February 6, 1992.

18. *Sunday Times*, February 2, 1992.

19. *Star*, June 11, 1991.

20. *Star,* January 28, 1992.

21. *Star*, February 8, 1991.

22. *Weekly Mail*, February 28, 1992. If there were such a direct, overt intervention by the SADF, a tradition of noncooperation would be mobilized to make the country ungovernable. We have a powerful tradition of civilian-based defense or defiance in South Africa to which the mass boycotts and strikes of the eighties attest.

23. Quoted in M. Ramushwana, "The Future of the Armed Forces of Transkei, Bophuthatswana, Venda and Ciskei: Expectations and Prospects," DISA Conference Paper, 1991, p. 6.

24. Quoted in the *Weekly Mail*, July 5, 1991.

25. Ibid.

26. *Sowetan*, October 15, 1991.

27. Nathan, "Riding the Tiger."

27. Nathan, "Riding the Tiger."

28. R. Williams, "Back to the Barracks: The SADF and the Dynamics of Transformation," *Southern African Perspectives*, Working Paper Series No. 10 (Bellville, South Africa: Centre for Southern African Studies, 1991), p. 28.

29. Quoted in the *Star*, January 28, 1992.

30. R. Williams, "Taming the SADF."

31. Nathan, "Riding the Tiger," p. 16.

32. Editorial, *Sunday Star*, June 3, 1990.

33. Quoted in the *Star*, October 24, 1990.

34. The possibility of setting up a regional conference along the lines of the Conference for Security and Cooperation in Europe (CSCE) model was suggested by Chester Crocker in February 1991. He argued that such a forum could grow out of the joint commission monitoring the Namibian/Angolan peace accords. The idea has subsequently been suggested by a diverse range of politicians, from President de Klerk to Carlos Cardosa, a former Mozambican government figure at the January 1992 meeting of the Southern African Development Co-ordination Conference (SADCC).

35. There is a strong argument that this should take the form of nonmilitary service in order to erode the legacy of militarism—the acceptance of violence as a legitimate solution to conflict—which is widespread among our youth. The ANC has consistently opposed military conscription. However, this view is shifting with the realization that conscript armies are less liable to independent political intervention.

36. Cilliers, "Integrating the Military," p. 3.

37. Nathan, "Riding the Tiger."

38. S. Baynham, "Security Strategies for a Future South Africa," *Journal of Modern African Studies* 28, no. 3 (1990), p. 419.

10

Reconstructing Regional Dignity: South Africa and Southern Africa

Peter Vale

Interstate relations in Southern Africa will be profoundly affected by how South Africa conducts its foreign policy; this is nothing new.

What is new, however, is that the region faces a series of daunting challenges in the 1990s. As global change has stripped away the patina of orderliness in other parts of the world, the ending of apartheid has sparked an interest in a range of issues that touch upon the future of both the region and its individual countries. These have not superseded residual concerns; rather, old anxieties have been intensified by the quickly shifting sands of international relations.

In this mélange of issues old and new, South Africa, strictly speaking, is an unknown quantity. Its regional policy, like much of its national life, is the prisoner of its incomplete domestic transition; as a result, policy is wholly unformed.

But the very newness of South Africa in the region opens the space for creative and farsighted initiatives. If South Africa is to help Southern Africa meet the challenges of our times, it will need to transform its international affairs toward universal goals of dignity, development, and democracy. This chapter describes both residual and new concerns in the region and sets out a series of challenges to South Africans as they restructure their regional policy process.

South Africa and the Region

Southern Africa's tangled and violent history has encrusted dependencies that bind all its people in a common cause. Against this backdrop, South

Africans are no more and no less than their regional cousins: they inhabit an increasingly forgotten corner of a changing world.

This common destiny, however, hides current realities. For worse rather than better, Southern Africa is divided into states. Its future therefore hangs on the outcome of interstate bargaining; this too is not new.

If history were a linear process, this would be easy to grasp. After all, the ruin South Africa has generated in the region is everywhere manifest. As this is written, the continuing war in Angola and the mutilated children in Mozambique are just two examples of apartheid's wanton regional legacy.

The ending of the Cold War, however, has shown that history advances by meltdown; sudden thaws bring to the surface deeply embedded issues. As a consequence, agendas are constantly restructuring. Recent Southern African events confirm this pattern, and South Africans who will make policy will have to recognize that the regional inventory is cluttered with compelling issues—some recurring, some novel—all requiring a creative and a measured response.

Of immediate importance for the region's security is the future of the conflict within South Africa itself. Unless South Africans discover the will to stop their civil war, there will be no peace in the region. More ominously, each upward ratchet in the contest is transmitted deep into the neighborhood. Despite efforts to seal borders, South Africa's internal discord continues to spew across the region's borders. In the 1990s, this spillage has taken on new and very dangerous forms because the region's borders have become, if anything, more porous.

The struggle for South Africa is especially destructive because of the high level of small arms within the country and the inability to police them. In 1991, for example, 11,577 firearms were stolen in South Africa; almost a quarter of these were taken from private homes. In the same year, a total of 197,509 applications were received for firearm licenses: of these, 18,268 applicants were refused registration.[1]

Anxiety is added to the fear of South Africa's closest neighbors by the recognition that the battle for the soul of South Africa is not only racial, but as Colin Bundy points in Chapter 4, it is also generational: a large young black majority faces an aging white establishment. Finding a less violent means of resolving the conflict within the country is consequently all the more difficult.

But there are other misgivings about South Africa in its neighborhood. Under the weight of its multiple strains, South Africa may crack, even disintegrate. This is not a wish but, in today's world, a sensible observation. After all, if the Soviet Union and Yugoslavia can disintegrate, can other multiethnic, highly armed states be far behind?

There will therefore be no regional peace until South Africans themselves find accord. But to underscore the point, this will not be easy. In no place on earth (with the possible exception of the former Soviet Union) is the transition to a post–Cold War regime likely to be more complex than in the Republic of South Africa.

In this, there is very little that those who make foreign policy can do. After all, the forces that formally drive South Africa's current violence are divorced from external factors; this was not always the case. In the dark days of full-blown apartheid, foreign policy, especially in the region, was nearly indistinguishable from domestic politics.

The prospect of South Africa's disintegration does raise different sets of foreign policy concerns. Not least of these is that each sovereign corner of a dismembered state, as Yugoslavia and the Soviet Union clearly show, begins to conduct separate foreign policies. If this happens to South Africa, the regional situation will be fraught with uncertainties.

Paralleling the Soviet experience, deep anguish will follow from the fact that South Africa has a potent war-making capacity. Its army is well equipped and highly mobile; though plagued with aging equipment, its air force is the region's finest; and no other country in the region has a navy. As important for the neighborhood is South Africa's indigenous arms industry, no match for the big powers to be sure, but impressive in African terms. As Solly Nkiwane has pointed out, "In the period 1985–1989, South Africa's total military expenditure was $17.109 billion, while the other states of the region, excluding Namibia, together spent $7.185 million as military expenditures. [These latter arms] . . . came from outside the region, in fact from outside Africa. South Africa's imports of arms for the same period totalled only $185 million—about 1.2% of its total military expenditure."[2] Add to this the country's self-confessed nuclear capabilities,[3] and it is easy to see why South Africa's neighbors consider the country a powder keg.

All this suggests that orthodox—no, near old-time—strategic concerns will dominate the regional horizon until South Africa itself comes to peace and until, by whatever means, it can assure its neighbors of its peaceful intentions. Until then, the states of the region will regard South Africa as a cocked gun.

Until this perception changes, there can be no serious economic development in the region. Sustainable growth, as events in Southeast Asia so graphically show, follows upon political stability. In most ways, then, the region is still in South Africa's grip. The ending of the Cold War may well have primed Southern Africa for liberation, but until the region's rogue is brought to heel, the states of the region will remain wary of South Africa.

Notwithstanding South Africa's gravitational pull, Southern Africa does have a political axis that is independent of the region's preponderant power. As a result, those who make regional policy in Pretoria will have to respond to a panoply of issues over which they ostensibly have little or no control. As servants of the regional power, however, they will need to appreciate that only cautious and honest behavior on these issues can help restore the region's confidence in South Africa.

Take, for instance, the deep shifts underway in the region's political geomorphology. Some of these, like the changes in South Africa, are the product of wider global change. Whatever their individual roots, they are likely to be disruptive to regional relations. How should Pretoria respond?

Some important lessons are to be learned from the changes currently underway in Zaire and Malawi. Both were sparked by deep shifts in geopolitical circumstances. In Zaire's case, the crumbling fortunes of Mobuto Sese Seko are in direct proportion to the United States' waning interest in Africa. Hastings Kamuzu Banda's fate in Malawi appears linked more to his long association with South Africa than to his age. But history may well prove that political change in South Africa helped to devalue Malawi's strategic importance to apartheid. Elsewhere in the region, too, change is underway—in Swaziland, Tanzania, Mozambique, Angola, among others. While each development will be measured, sudden shifts in the geology cannot be discounted. Because of the unpredictability of the times, the region will be lucky if these transitions all turn out to be as smooth as Lesotho's. Without a shot fired in anger, a military that had itself overthrown a single-party government handed power to a political party after an election. In nothing short of a miracle, the once exiled Basuto Congress Party became a democratically elected one-party state!

For policymakers the lessons of change lie beyond the wider disruption it can cause. They must recognize that the political transformations of the kind experienced in Lesotho are very different from those of the independence era of the 1960s and 1970s. In those halcyon days, liberation gave states political space even if only to play the superpowers off against each other.

Today, the move to democracy effectively restricts the capacity of both governments and leaders. Only the stouthearted would want to share the fate of Zambia's Frederick Chiluba, a democratically elected leader of a country with soaring unemployment, poor terms of trade, depressed commodity prices, and eager young free marketers of the international financial institutions beating a path to the door of Government House.

Even democratically elected governments can become embroiled in conflicts with their neighbors. South Africa's policymakers will have to understand the manifold sensitivities around conflicts of this kind. Managing these will be all the more difficult because the region has a poorly developed tradition of conflict resolution, despite the fact that the protocols for regional integration are fairly well developed.

From a comparative perspective, the Southern African Development Conference (SADC, formerly SADCC) has an interesting genealogy. Most attempts at regional integration, even of an economic kind, aim to surrender sovereignty. SADCC was a creature of a different kind: it was primarily aimed at defending the sovereignty of individual members from South African aggression. At the August 1992 Windhoek meeting, which transformed the development coordinating conference into a regional community, security aspects were added to SADC's routine business, but it remains to be seen whether this will encourage the pacific settlement of disputes between neighbors.

As events play out in the region, mechanisms of this kind will increasingly be needed since many of the region's borders are under dispute. The most obvious example is the South African port and enclave of Walvis Bay on the Namibian coast, a case that has been handled by the two neighbors with much equanimity. Less certain is the outlook for the resolution of other border disputes, as the case of the conflict between Namibia and Botswana over the territory once known as Swampy Island demonstrates. The trauma of this particular case is highlighted by the fact that the two countries call the island different names: to Namibians it is Kasikili Island; to Botswana it is Sidudu Island.[4] Tiyanjana Maluwa points out both the Monty Pythonesque and the serious face of the conflict by observing that to

> the casual observer, the spectacle of two apparently friendly neighbors threatening to fall out over contested sovereignty over an uninhabited island which is, in any case, submerged under water during a certain period of the year seemed somewhat ludicrous, if not farcical. . . . [But] the importance attached by nations to territorial ownership needs little demonstration or emphasis.[5]

In Southern Africa the propensity of border tensions to quickly escalate appears high because of the cloak of stability that apartheid purportedly drew over regional relations. The absence of an established mechanism—like a regional court of justice along the European model—to deal with these conflicts will make them difficult to resolve.

But sound regional relations are not only for the impervious, that is, the states that can match South Africa's own bureaucratic machinations. Weaker states of the region too will have to be carefully monitored because of their very vulnerability. Global change and the region's uncertainties will worsen their sense of exposure.[6] As seriously, weak states are exposed to the predatory ambitions of large neighbors. Policymakers will have to be especially vigilant in determining South Africa's relations with these states.

At each level of the debate on the region's painful need to reconstruct itself in the face of wider change, new issues emerge. As South Africa's policymakers grapple with these issues, they will have to understand that policy, like life, can be understood backwards but must be made forwards. To be constructive, coherent, and competent, sound regional policy must look toward the future and anticipate issues that can suddenly ascend the policy agenda.

If the region is to reduce its dependency on foreign aid, as many believe it must, serious policy adjustments will weaken the stability of governments and further weaken individual states. In some cases, this will become linked to the idea that sovereignty—as we now understand the concept—may have to be redefined. This is really a Southern African version of the global debate. But its implications will be profoundly unsettling because sovereignty has been the central organizing concept in the development of the subcontinental state system. Moreover, in the region, as elsewhere in Africa, sovereignty has come to enjoy a far stronger currency as a result of the famous injunction by the Organization of African Unity (OAU) that African states should respect the authority of colonial boundaries.

If the centerpiece of interstate relations is under threat, then the entire system of conflict and its resolution may change its cadence. In this situation, states may fracture and disintegrate. This, too, will occur in Southern Africa. Consider the case of Tanzania: if—as the present direction of that country's discourse suggests—the union collapses and Zanzibar is launched on a separate trajectory, then that island will become susceptible to all the security problems faced by small states.

It is instructive to recall, however, that predicting secessionist movements in the Third World is a graveyard for analysts. In the 1960s and the 1970s, there was only one successful secessionist, Bangladesh. The bloody conflict over Biafra shows how a state, with the willing support of allies, can coerce its citizens into submission. What this means in the 1990s is still uncertain; plausibly it might be argued that the greater concessions toward ethnicity make it inevitable that fragmentation might become the rule of international life.

The long-term predicament for policymakers is to manage a process by which threats to the sovereignty of the region's states can be met without generating panic and new forms of instability. Trauma is added to this by the recognition that the OAU, the one African body charged with regulating conflict, is itself profoundly threatened by sovereignty's waning hold. All this is not conducive to building internal cohesion nor, indeed, wider international confidence in the continent.

These general points on the implications of political change in Southern Africa are augmented by a range of more practical concerns. Each of these, as South Africans are discovering, is to be found at the global level.

One is the question of natural resources. The devastating drought of the 1991–1992 season, which deeply touched security relations, has highlighted a regional soft spot: without water, southern Africa is a dust bowl. This is one reason why South Africa's policymakers will come to recognize (perhaps, even value) some decisions taken by South Africa in the 1980s. The links between South Africa and Lesotho, which are known by the generic phrase the Lesotho Highland Water Scheme, were politically conceived but were grounded by longer-term considerations of resource scarcity.

The region's vulnerable food basket is another potentially serious source of tension. Policymakers will need to understand why it is that overall agricultural production declines in a region that, when the rains do fall, is very fecund. They will need to master the emerging series of explanations that link the decline in food output to poor international prices for agricultural commodities; the life and death nature of the issue means it can no longer be simply ignored.[7] The immediate multilateral challenge is to develop mechanisms that will foster cooperation on food security.[8]

The rediscovery in the 1990s of the importance of life-providing commodities is linked to the deepening concern for the environment. But the devastating regional drought of 1992 has been primarily ascribed to changing weather patterns. Although there is a nascent debate on this issue in, particularly, South Africa, it is totally underplayed in the regional discourse. Almost certainly, Southern African governments will be compelled to pay more and more attention to the issue. This will be difficult—perhaps, impossible—since relatively sophisticated responses need to be put into place, and often these destroy distinctive features of national life.

How, as an example, does an African state deal with issues of overgrazing, which contributes to desertification, when cattle populations have grown dramatically? How do they deal with degradation of soil as the human load on the same soil increases? And how do governments counter deforestation in the face of population increases? These are not the only threats to Africa's environment. Air pollution, acid deposition, groundwater depletion, pesticide and heavy metal contamination, environmental diseases, and (as we have already seen) climatic change are some of the many possible environmental problems that could significantly influence Africa's long-term future.[9]

The sheer physical capacity of the region to carry ever increasing numbers of people is an important question that impinges directly on policymaking in the field of foreign affairs. While Malthusian concerns of population questions have been modulated by advanced understanding of the carrying capacity of the planet, few can believe that Southern Africa's average population growth rate of 3.5 percent cannot have serious effects on the entire region from a number of points of view. Two effects, for example—cross-border migration and the future of the region's cities—will demand the attention of South Africa's policymakers.

The pathology of migration in southern Africa is no different from that in other parts of the globe. A series of linked circumstances have set in motion a tide of humanity across the face of the subcontinent. In turn, this has impacted upon the region's well-established patterns of migrant labor. For centuries, southern Africa's people—sometimes on a seasonal basis—have crossed its borders in pursuit of work. In most cases this flow has been southward toward the Witwatersrand, the region's industrial heartland. The ending of apartheid appears to have unleashed a tide of migrants, who see South Africa as a suitable location in which to better life's chances.

Southern African migration resembles, therefore, the kinds of patterns the Europeans are currently experiencing.[10] The threat to intercommunal harmony that migration has posed in Europe might conceivably be repeated in Southern Africa.[11]

Waves of migration can disturb tolerable patterns of social relations within host countries. Certainly, judging by the numbers of immigrants in small business in South Africa, it seems probable that they will transform this sector of the country's economy. This should not be surprising since, to make a general point, immigrants are likely to be more educated than South Africans who have suffered generations of deprivation as a result of apartheid. If they prosper in the small business sector, they will become better off than South Africans.

New patterns of economic and political dislocation throughout the region have installed immigration and its cousin, the refugee condition, as almost permanent features of regional life. Almost every country in the region is a home or a host country for migrants. Continuing strife in Mozambique and Angola has scattered refugees across the east and the west of the subcontinent. But the myriad of issues associated with both drought and debt have further scattered the region's people.

Aside from the intrinsically destabilizing nature of this movement of people both for the host and the home countries, the subterranean features of mass migration are increasingly treacherous: the killer diseases—AIDS, malaria, TB—cross borders at an alarming rate. As cross-border migration increases, it also becomes difficult, if not impossible, to control the flow of additional migrant-borne threats: small arms and drugs are just two obvious examples.

Africa is no longer a rural continent. Indeed, the drift to the cities of the last decade has ended the myth of the predominance of the rural-based African peasant.[12] Within the continent, southern and northern Africa are the most urbanized, and western and eastern Africa the least. The push toward the cities means far more concentrated populations and the real prospect that the region's urban concentrations will become megacities. Exact numbers are difficult to estimate, but cities in excess of 5 million people become very difficult to manage in a First World setting. In Southern Africa, they will be near impossible because its cities too were built for the comfort of the ruling minorities.[13]

At first blush these issues might seem far removed from the foreign policy agenda. However, the weakening capacity of the state in Africa holds out the prospect that with time its cities can overtake the state itself. As the European experience teaches, cities can become independent foreign policy players; at times they can compete with the international goals of central governments.

Those who make South Africa's external policy will have to be as close to the politics of this as they are sensitive to the problems of unprecedented urban growth. The reason is clear: the political agglomeration of a weakening state, massive urbanization, and economic stagnation will have far-reaching effects on a range of issues upon which this chapter has already touched—"fertility, employment, education, agriculture, political stability, and other demographic and socio-economic factors."[14]

The long-term prospects for Southern Africa will follow from the region's capacity to balance different sets of relations. In all these, of course, South Africa is preponderant. The challenge for South African policymakers is simple: Can they avoid regional conduct that is instinctively bellicose? If they cannot, they will remain the region's bully. If they can, how will they engineer the policy process? Given the region's weight, can South Africa help to ensure harmony in one of the world's most troubled regions?

Restructuring the Policy Process

The Chinese curse "may you live in interesting times" has a double-edged meaning for those who will govern South Africa when apartheid's walls all finally crumble. Global change has left diplomats and academics alike grappling to find new approaches—even perhaps a new lexicon—for describing international developments. This problem of understanding is compounded by the intellectual Somalia syndrome: although there are many good ideas around, powerful conceptual warlords continue to protect increasingly fragile conceptual fiefdoms.

On the one hand, then, South African policymakers will have to face the challenge of new realities and accept that there are only limited passages to understanding and appreciating their dynamic.

On the other hand, they must wrestle with the daunting task of creating a new self-image of their country. Until this happens, South Africa's official international relations will be simply an extension of what has gone before—a set of positions informed by narrow understandings of national interest, driven by unacceptable patterns of international behavior, and largely out of touch with the hopes and desires of the country's people.

Only a new, distinctive imprint of South Africa as a fully accepted international player can help drive a creative new regional and foreign policy. Establishing this will not, however, be easy; like its domestic policy, South Africa's foreign relations were blighted by apartheid.

Until its political process was freed, the country conducted not one but a series of foreign policies. A powerful one was located in the apparatus of the state and essentially had two dimensions: a security arm that, among other things, caused havoc in Southern Africa, and a diplomatic arm. This was the foreign policy of South Africa's minority rulers.

The liberation movements conducted an equally potent strain of foreign policy. Nurtured in the exile experience, its public persona aimed at isolating the minority-ruled state while its military dimension sustained efforts to destroy apartheid by force of arms.

These two traditions have recently been dubbed the "upstairs" of state diplomacy and the "downstairs" of the diplomacy of the liberation movements. Although some have called these terms pejorative, they are useful analytical bridges because they explain the dualistic nature of South Africa's international experience. Like so many other sectors of its society, the upstairs and the downstairs were formed and informed by discrete understandings of South Africa's realities; each perspective is a prisoner of a certain understanding of the country's history.

The ending of the Cold War enabled both to accept the limitations of their own positioning in the world. So, South Africa's minority recognized that it was not a specially cloistered, racially pure enclave that was to be defended at all costs because it safeguarded Western interests in a hostile corner of the planet. And South Africa's majority, among other things, came to realize that the Soviet commitment to their cause was weakening in parallel to the deepening crisis in Moscow.

If, however, South Africa is to emerge as a serious international player, it will have to meld together its two experiences of the world. This will not be easy because each tradition has different instincts and different understandings of priorities. These have come sharply to the fore in the nascent debate over South Africa's economic position in the emerging world order.

The upstairs school believes that South Africa has no option but to carve a niche for itself between NAFTA, the European Community, and the Asian countries: a triangle of trading blocs called the Triad. This crudely mercantilist view holds that middle-ranking countries with limited strategic purchase, like South Africa, will have little or no choice but to fall into line with the economic designs of the Triad.

The downstairs sees things differently. The emerging world order, they hold, will not fully serve South Africa's interests. They consequently recommend that the country contest the rules as they emerge. This is to be achieved by seeking a series of links with states and formations with the countries of the South. The nub of their position is that South Africa's global links should serve its own interests, not those of states that would ultimately marginalize it. This is not altruism, they argue, but a position spurred on by collective self-interest.

These strongly divergent views follow from different understandings of the current situation and are formed by different experiences of the world. One measure of the country's capacity to forge a new self-image will be how it manages the transformation of the state agency responsible for regional and international issues, the Department of Foreign Affairs.

Measured by any objective standard, South Africa has not had a professional diplomatic service. True, the formal (read the state's) foreign service has worked long and hard at developing a professional ethos. Their efforts, however, were thwarted by the apartheid question. This bald statement needs qualification on two counts. First, a group of diplomats serving South Africa did try to set themselves above the political fray.[15] It is fair to say, however, that the apartheid weight increasingly drew them from this goal. Second, at a functional level South Africa's diplomats were required to perform day-to-day diplomatic and consular work; by all accounts, this was proficiently done.

The history of South Africa's international positioning produced, as we have noted, another dimension: apartheid's opponents also conducted a species of foreign policy. In the West, this work was a combination of political activism and information offices; in the former East and in Africa, it was combined with regular diplomatic work. But—and this is the important point—it served specific political objectives: it aimed at the isolation of the apartheid state. In no way can it be considered professional diplomatic work.

Driven partly by economic considerations and partly by the wider international changes, South Africa is set to integrate its two experiences of the world. It is destined to create from two foreign policy cultures a single one, and bureaucrats will be the tools of this experience.

This is nearly unique in modern times. The Eastern European experience aside, two other efforts at changing foreign policy culture, Namibia and Zimbabwe, were somewhat different from what South Africa faces. In Namibia's case, a new service was established; since Pretoria had handled the foreign dimension of its ward's interests, the end of South Africa's occupation gave the Namibians the opportunity to initiate a new service.

While Zimbabwe's experience is often thought to be analogous to South Africa's, it is in fact very different. In the former Rhodesia, the twenty-year sanctions experience reduced the foreign office to a shadow. Moreover, the prevailing ideological mood when Zimbabwe gained independence made it almost inevitable that the Smith regime's diplomats be—largely because of the Cold War—of very little interest to Robert Mugabe and his colleagues as they entered government.

The restructuring of state departments is not easy, and South Africans will have to tackle it conscious of looming financial constraints and, as we have noted, in the face of global change.

In many important ways, South Africa's path to developing a new self-image will be smoothed by the impressive range of second-track diplomatic instruments close at hand. The country has world-class universities, a potent trade union sector, an energetic business community, churches deeply steeped in international relations, thriving NGOs, and parastatals well versed in regional affairs. Also, individual South African names—Tutu, Mandela, de Klerk, to mention three of perhaps a thousand—are household names across the planet.

But sound backdoor diplomacy is no substitute for robust international engagement and, as we have suggested, this can develop only after South Africa fully understands its role in the international system—until its image of itself in the world mirrors its understanding of itself as a country. All this, as we have learned, will not be easy. And it is in Southern Africa that the new South Africa's true test as a responsible international citizen will take place. In the region, South Africa's new foreign relations will be most clearly reflected, and here too patterns of behavior will be shaped and reshaped.

South Africa's search for a new self-image that will drive its foreign policy will be reinforced by the series of organic processes already underway throughout the region. Notwithstanding its regional preponderance, South Africa cannot halt the region's progress; but it is equally true that Southern Africa's sense of destiny will be hollow without South Africa.

Global change has profoundly altered the ways in which Southern Africa looks toward its future. In the old days, the Good Guys all wore white hats, stood for the correct things, and supported each others' views of the world; the Bad Guys wore black hats, stood for the wrong things, and supported the other side. This parody of human and international behavior no longer plays, however. Consequently, the choices states must make are far more difficult. South Africa's policymakers in particular will discover that they are caught between increasingly narrow options: between the rock of more issues to think about and the hard place of shrinking budgets, as Jeffrey Herbst has shown in Chapter 3.

This inventory of challenges for South Africa's policymakers is intimidating. Many have understandably asked whether, given its immense problems of domestic reconstruction, South Africa will have the time, let alone the resources, to deal with these. Ironically, both South Africa's long isolation and the processes of global change teach the same lesson: no country can afford to be excluded from the international community. More prosaically put: these days, if a country has a border and a fax machine, it needs a foreign policy.

Can South Africa lift the payload the international community expects of it? No one can say for sure. As they go about their demanding business, policymakers might draw comfort from the fact that the international community isolated apartheid, not South Africa's people.

Support from the wider international community will certainly help South Africa develop its self-image within the wider world. But the real test is for South Africans themselves. Will they, especially in the neighborhood, draw the lesson from Cicero's question: "Where is there dignity unless there is honesty?"

Notes

1. *Paratus* 43, no. 5 (May 1992), p. 8.

2. *Southern Africa: Political and Economic Monthly* 6, no. 2 (November 1992), p. 10.

3. *Washington Post*, May 1, 1993.

4. This is a small uninhabited island in the Choebe River located within the area bounded by, approximately, 25° 07' and 25° 08' E longitude and 17° 47' and 17° 50' S latitude. See Tiyanjana Maluwa, "Disputed Sovereignty over Sidudu (or Kasikili) Island (Botswana-Namibia)," in *Southern Africa: Political and Economic Monthly* 6, no. 2 (November 1992), pp. 18–22.

5. Ibid., p. 18.

6. C. E. Diggines, "The Problems of Small States," *Round Table*, no. 295 (1985), pp. 193–194.

7. For a discussion of this, see Carol B. Thompson, *Harvests Under Fire: Regional Co-operation for Food Security in Southern Africa* (London: Zed Books, 1991).

8. See Peter Vale, "A Drought Blind to the Horrors of War (. . . and the Challenge of Peace)," *Die Suid-Afrikaan*, August–September, 1992, pp. 51–52, 57.

9. Chinua Achebe et al., *Beyond Hunger in Africa: Africa 2057: An African Vision* (London, James Currey, 1990), pp. 70–71.

10. See François Heisbourg, "Population Movements in Post–Cold War Europe," *Survival* 33 (1991), pp. 31–43.

11. Gil Loescher, *Refugee Movements and International Security*, Adelphi Papers No. 268 (London: International Institute of Strategic Studies, 1992).

12. This point underpins one of the most interesting books on Africa to come out in recent years; see Basil Davidson, *The Black Man's Burden: Africa and the Curse of the Nation State* (London: James Currey, 1992).

13. In early January 1993, South Africa's Sunday press reported on conditions in the Johannesburg suburb of Bertrams. One story claimed that sixty-three people were living in a house in Ascot Road (*Sunday Times*, Johannesburg, January

31, 1993). Another report claimed that sixty people were occupying houses in the same suburb (*Sunday Star*, Johannesburg, January 31, 1993).

14. Achebe et al., *Beyond Hunger in Africa,* p. 45.

15. South African diplomats have been reluctant to write up their stories. A rare—and therefore interesting—exception is Donald Sole, *"This Above All":* *Reminiscences of a South African Diplomat* [no place, no publisher] 1990.

11

Southern Africa's Transitions: Prospects for Regional Security

Gilbert M. Khadiagala

Discussions of regional security dominate most analyses of Southern Africa. These analyses are informed by the need to explain the impact and relevance of ongoing transformations on the future of regional relationships. Prevailing transitions, however, make predictions difficult; the past is dying without giving rise to a manageable present or a predictable future.

Since they involve some discernible movement from one phase to another, transitional periods always contain the potential for projecting new issues and actors into the policymaking arena. Thus, the resolution of one issue might conceivably lessen the significance of actors associated with it; other actors previously marginal to the process might then take their turn at decisionmaking. New security problems often, therefore, require different actors with distinct approaches and calculations; in a word, transitions could cause discontinuity in perceptions of security. By the same token, transitions could be decidedly deceptive events, promising changes in actors and approaches yet concealing continuities. Moreover, when old actors remain on the scene to confront emerging security issues with the same policy arsenals, transitions become no more than breathing spaces, akin to operatic interludes. Despite their illusory nature, however, transitions still need to be studied because they potentially represent conspicuous departures from a defined past and yearnings and fears about a new future.

Present transitions in Southern Africa provide its states with an opportune moment for redefining modes of interstate cooperation. Although the inconclusive nature of national and regional transitions presents problems for predicting a clear pattern of regional security in the short and medium terms, long-term scenarios point to the establishment of regional problem-solving mechanisms along more minimalist lines. I argue that while such minimalist structures may mean the postponement of a dream of a fully integrated regional community, they are appropriate for the postapartheid era. As states address the long-standing issues of domestic insecurity and work to establish viable democracies, the embrace of functional cooperation will gain legitimacy.

Retrospective Views of Regional Security

Between 1960 and 1990, regional security in Southern Africa revolved around decolonization conflicts. A remarkable subsystem shaped by geography and colonialism, it witnessed both the centrifugal forces of wars of national liberation and the centripetal forces of regional economic interdependence pushing in opposite directions. With its constituent units consumed by these conflicts, its security assumed explicitly external dimensions.

More so than for states in the wider continental system, security for Southern African states was intimately tied to the purpose of realizing the ideals of African nationhood and dignity. The obduracy of colonial and racist regimes only strengthened the links between security and decolonization. Throughout this period, the nature of threats to the region's independent states was not some imaginable danger to sovereignty, but the more tangible one of survival within a malignant environment. Elite strategies of various shades—from the extremes of Samora Machel's Mozambique to Kamuzu Banda's Malawi—largely reflected a conscious need to adapt to overweening minority regimes in the regional neighborhood.

At the height of decolonization, the achievement of nationhood was just a first step toward the broader goal of confronting a hostile regional environment; the road from the traumas of nationalist struggles led invariably to the realities of economic dependence and, increasingly, to South African destabilization. Yet interdependence in adversity contributed to the creation of equally tangible attempts at regional cooperation and problem solving. Hesitant at first, these attempts nevertheless culminated in the establishment of the Frontline States (FLS) and the Southern African Development Coordination Conference (SADCC) as the minimal cooperative mechanisms for regional underdogs.

As most writers have shown, these alliances contained both the strengths and weaknesses of subordinate states to pursue relatively grandiose goals on limited resources.[1] Aggregating strengths to compensate for their weaknesses allowed them a measure of leverage in regional affairs but was not sufficient to meet all of their objectives. Given this quandary, most policy outputs of these alliances exhibited what Sam Nolutshungu refers to as "intentional ambiguity."[2] At the regional level this ambiguity was evident in efforts by some states to accommodate the interests of minority regimes even while agitating for their demise; for instance, Zambia at one point succeeded in making the objective of Zimbabwe's decolonization entirely compatible with reopening transport routes through that country. At a more critical level, SADCC largely thrived on reconciling the long-term goals of reducing dependence on South Africa while increasing this dependence in the short term. Put succinctly, intentional ambiguity was a markedly successfully strategy for carrying both the burdens of consolidating nationhood and contributing to decolonization.

A large part of the conceptualization of regional security in the past revolved around the issues of external participation in the subsystem. Whether it was the extent of the communist threat to Western interests, the "communist onslaught" on South Africa, the imperialist threat to national liberation struggles, or simply the levels of aid SADCC partners would disburse in a given year, the debates on foreign involvement underscored the differentiated nature of external actors in the region.

Here too the regional alliances pursued a variety of ambivalent approaches: they simultaneously requested arms for the liberation struggle from the former Soviet bloc nations while demanding economic assistance and mediation efforts from Western countries. This posture toward extraregional actors produced a number of results. In just under ten years, external actors had helped to decolonize Zimbabwe and Namibia, secured the Angolan regime, and mobilized a measure of economic and military support for the beleaguered Mozambican regime. Whereas the FLS decried superpower intervention, the conflicts they engendered provided the impulse that generated systemic changes, allowing regional states to borrow power at a relatively low cost.[3] In addition, foreign assistance was largely instrumental in providing the necessary capital that gave SADCC its present regional standing. Strengthening the momentum for cooperation, SADCC enabled the region's states to withstand increased pressure in the era of destabilization; to the extent that it stymied South African efforts toward constellation, foreign assistance helped tame the regional leviathan.

The conclusion of decolonization resolves one of the major sources of regional insecurity; with it, however, come issues of finding new modes of interstate relations consistent with the emerging priorities of domestic reconstruction. The present regional context for the pursuit of these objectives is constantly changing, presenting both opportunities and challenges to decisionmakers.

Prospective Views

Internal Dimensions

Attempts to define new regional relationships that will conform to global changes hinge, more than ever before, on the outcomes of domestic transitions. The broader focus of these transitions is to redress the domestic sources of insecurity by increasing political participation and restoring the foundations of nationhood. But one of the major uncertainties revolves around whether elites used to the status quo are capable of making compromises that are essential to enduring processes of institution building and economic recovery.

Across the region, demand for multiparty democracy promises to unleash new conflicts and put forth new leaders. The collapse of communism and the stagnation of regional political economies have already produced a period of potential reorientation throughout the region. Reflecting on these changes, John Saul has perceptively observed that "there is now that much less conceptual ground on which to hide opportunistic leaderships of a Zambia or Zimbabwe who seek the solace of one-party solutions to political complexity in their countries."[4]

There are two broad categories of internal efforts directed at revitalizing Southern African political systems. The first consists of the negotiated transitions in Angola, Mozambique, and South Africa, where ruling parties entered into talks to break long-drawn political and military stalemates. In Angola, the MPLA's triumph in parliamentary elections of 1992 seemed to offer it the mandate it has sought over the years, but this mandate was denied by UNITA's decision to return to seeking power with the gun. Without peace, it is inconceivable that the attempts by Angolan leaders to create a thriving Angolan nation and economy will succeed. In Mozambique, two years of negotiations between Frelimo and Renamo resulted in the Rome agreement for a comprehensive cease-fire that would lead to elections in 1994; what will remain unclear in the immediate future is the commitment of both parties to a broad-based process of reconciliation and confidence building as they enter the uncertain terrain of electoral competition. The crippling effects of war on the economy will ensure that even if a political transition is successful, Mozambique must then grapple with the Herculean task of economic reconstruction. In South Africa, three years of talks have finally produced arrangements for a transition to formal democracy. But as analysts in this book warn, obstacles ranging from security and right-wing forces who want to derail democratization and majority rule to continuing township violence and the need to transform the economy and local government and address the concerns of youth and women all point to the protracted nature of South Africa's break from the past.

The second category of political change is the transition of dominant African one-party systems into pluralist systems. The multiparty transition in Namibia gave impetus to moves toward democratization in Lesotho, Malawi, Tanzania, and Zambia. The Namibian transition was also instructive in dissuading Zimbabwe's Robert Mugabe from the folly of making his country a de jure one-party state. More important, the ouster of Kenneth Kaunda from Zambia's presidency through electoral mechanisms emboldened budding opposition movements across the region. Underlining this renewed confidence in democratization, Frederick Chiluba, the new Zambian leader, observed that the stream of democracy "dammed up for 27 years has been finally freed to run its course as a mighty African river."[5]

Overall, the most profound questions raised by these domestic changes center on whether pluralism and expansion of political space will lead to genuine national renewal and transformation or merely fragmentation along ethnic, class, and regional fault lines. Might the newly unleashed "civil societies" overwhelm the economic and political capacities of "reforming" politicians? Does the democratization process have structural roots in Southern Africa or is it merely a palliative to address the ephemeral problems of political economies under siege? Drawing on the experiences of the immediate postcolonial period in the rest of Africa, some scholars have warned of the euphoria accompanying the democratization processes; they caution that building democracies on culturally diverse bases in the absence of elite consensus is bound to be a frustrating task. Further, after passing through the preliminary phases of democratic transitions, these countries confront daunting problems of transforming their economies and building participatory democratic cultures and mechanisms.

Regional Dimensions

Old and new political elites alike have to make decisions relating to interim measures of regional cooperation before the transition processes sort themselves out. Domestic transformations dovetail, and are intimately linked, with changing perceptions and priorities of state actors about their place in the regional equation. Like the domestic arenas, the regional environment presents both opportunities and constraints; constraints lie in new forms of pressures from politically interventionist yet economically less charitable external actors. Opportunities, on the other hand, inhere in the possibilities of building bridges across the erstwhile racial divide.

With respect to the role of external actors, the conclusion of decolonization conflicts saw a marked diminution of external military intervention in regional conflicts. The lifting of the East-West political-military rivalry, has, nonetheless, been accompanied by enhanced Western economic intervention via the dominant Bretton Woods institutions, the World Bank and the International Monetary Fund (IMF). For the most part, Western hegemonic sway is symbolized by the political conditions the World Bank and IMF are imposing on economic aid as part of the crusade for accountability and democratization. Complaints about the deleterious impact of the World Bank's structural adjustment programs illuminate the predicament of reconciling dependence on external economic aid with that of maintaining a semblance of political sovereignty. The present phase that Thomas Callaghy aptly characterizes as "enhanced neocolonialism"[6] promises to curtail the room for maneuverability of Southern African states.

A related issue affecting regional actors in the post–Cold War era is the now familiar one of marginalization. While the consequences of marginalization are often projected in terms of the shifts of aid resources away from Africa to Eastern Europe, their more durable impact seems to emanate from obstacles to African market access subsequent to growing protectionism in the West. Internal dynamism in the European Community, captured in the Maastricht project, portends far greater consequences for Southern Africa than the decline in aggregate aid levels.

The opportunity for building bridges within the region is underscored in the wide array of economic and diplomatic links between the FLS and South Africa since 1990. Christopher Coker emphasizes the need for these openings largely because Southern Africa should forge a regional life and meaning of its own choosing: "So far, Southern Africa's meaning has been a construction of the outside world. . . . In the past, local states made too many demands on the outside world in terms of arms, sanctions or aid, [yet] too few of each other."[7] South Africa's diplomatic offensive in 1990, appropriately christened the "New Diplomacy," began as a counterpart to its domestic normalization process; it involved breaking out of regional, African, and international isolation. As Peter Vale observes, the underlying assumption of this policy was that

> success abroad would secure de Klerk's efforts at home: diplomatic recognition by South Africa's neighbors would carry forward the understanding that the incumbent government had the necessary authorization to dictate the pace and progress of domestic change. This would help, in addition, to speed the end of sanctions.[8]

South Africa's neighbors responded with cautious optimism to these overtures. Visible movement toward change in South Africa appeared to be the start of checking the cycle of violence and instability that had characterized the region for decades. Although some were reticent to readmit South Africa into the political fold due to concerns about alienating the dominant nationalist forces, the imperatives of normalization prevailed. Their precarious economic situation as a result of the structural adjustment programs and drought in the region made them more receptive to promises of trade and limited foreign aid from South Africa. It was against this background that initial promises to isolate Pretoria until there was a "profound and irreversible process of change" foundered on competitive unilateral moves by the FLS and SADCC states. The minimal level of consensus that had typified their approach to regional issues began to evanesce as most of them made unilateral initiatives, primarily toward economic cooperation with South Africa.

It is, however, in the area of regional institutions that the transition poses problems to African states. At the heart of this dilemma is the future of SADC (the former SADCC), the symbol of collective endeavor by the FLS to strengthen their structural position in the subsystem. Throughout the 1980s, SADCC was sustained by a tentative but powerful ideological rationale of reducing dependence on South Africa; more critically it became a tangible embodiment of their relative success in attracting international attention. Yet the peaceful regional environment characterized by the absence of pressure from Pretoria threatened SADCC's leverage to attract external support and deprived it of its exclusive position as an alternative avenue of economic access. Simba Makoni spoke to this dilemma by noting that although SADCC generated a lot of support since its formation,

> I regret to say that a large part of this support has really not come to us on our own account. It has come to us as sympathy support against apartheid. People saw our countries being destabilized, being aggressed, being attacked by South Africa, and this was their response. It was not their response to SADCC, and that is going to be our next major challenge in terms of our international relations. People have to accept SADCC for what it is, and we have a lot of positive things in our own right—SADCC as SADCC, with or without South Africa. . . . We have to generate support with or without apartheid. . . . That is both a challenge and a disappointment because there are a lot of people who have read us in the opposite image of apartheid.[9]

The decrease in SADCC's external appeal was compounded by misgivings over its relations with South Africa. During this transition phase, questions about South Africa's role were tied to its overall potential contribution to regional economic cooperation. Optimistic scenarios envisaged South Africa playing a benign role, commensurate, in de Klerk's words, with its status as "Africa's Japan."[10] Proponents of this scenario contend that as the dominant economy, South Africa will play its natural role of economic leadership; as Simon Brand, the head of the South African Development Bank, argues: "In a normalized environment in the region as a whole . . . South Africa will be able to play the normal kind of leadership role that its economic prominence in the region, the strength of its institutions, both in the public and in the private sector, really suggest that it should have played a long time ago already."[11]

The collapse of international consensus for sanctions revealed that even in the absence of a clear picture on an overall postapartheid package, South Africa could reestablish its image in the eyes of influential external actors. Douglas Anglin remarks that governments in the region "found themselves in double jeopardy since, as the political rehabilitation of South Africa proceeds, that country is once again becoming a powerful rival magnet for overseas investment and even aid."[12]

It is partially South Africa's growing international acceptance that leads pessimists to posit its hegemonic intentions within the rubric of any conceivable regional economic arrangement. Colin Stoneman and Carol Thompson, for instance, contend that international financial institutions and leading industrial powers seem to be increasingly in favor of regional integration with market forces determining trade and investment flows; the consequences of this approach, they note, "would primarily benefit the largest of the economies, South Africa, and leave little room for coordination of economic activities by SADCC governments or for national industrialization strategies."[13] Others have indicated that as it gains more respectability and acceptance, South Africa will draw countries such as Botswana, Lesotho, Swaziland, and Namibia further into its economic ambit to the detriment of more geographically distant regional states.

Fears of South African dominance persist in spite of the ANC's oft enunciated policy that a future democratic government would renounce all hegemonic ambitions by basing relations on principles of equality, balance, and mutual benefit.[14] Based on this perspective, many within the region believed that the best way to preempt South Africa's hegemony was to strengthen attempts to build a regional economic bloc. Toward this end, SADCC became SADC and launched a program of economic integration in January 1992 similar to the EEC that envisages provisos such as a single regional market; free movement of goods, labor, capital, and services; and a single currency by 1994. The establishment of SADC was predicated on the notion that weaker states should create workable programs and effective institutions that will allow the eventual integration of South Africa on terms more favorable to the region.

During this transition phase, the empirical process of unilateralism by South Africa's neighbors points to a possible realignment of regional relationships. Apart from the economic contacts that so far sidestep the institutional arrangements of SADC, unilateralism is evident in the stream of diplomatic and cultural contacts established between most regional states and South Africa.[15] Besides, South Africa's success in bridge building outside Southern Africa has bolstered its regional position by dramatizing its supposed extraregional capabilities; it is instructive to note that de Klerk has consistently invoked images of "regional powerhouses" revolving around Egyptian, Kenyan, Nigerian, and South African growth poles that would provide "stability, political progress, and economic growth for the continent as a whole."[16]

Fundamentally, SADC's short-term future has already been affected by the conscious questioning of its capacity and goals by the new regime in Zambia, underscoring again the link between domestic change and regional processes. Even before the transition in South Africa has taken a definite shape, Chiluba called for a much speedier economic reintegration of South Africa. Chiding the previous government's insincerity with respect to relations with South Africa, he noted:

> We intend to trade with South Africa . . . we want to take advantage of economies of scale in South Africa. In terms of the statistics available, trade between this country and South Africa has never declined. It is rather the hypocritical language which was applied by the previous administration that fooled the world into believing that there was any slow-down or decline in trade. There never was at all. What we intend to do is, in addition to the figures and facts that do exist, we want to speak a language that is compatible with the facts and figures, so that the world knows that we are neighbors and we trade together. . . . We do believe that whilst we are kind of helping in the political process, South Africa and the rest of the countries in the area must begin to think seriously about turning the whole area into an economic unit; integrated for common purposes.[17]

Apart from allowing South Africa to open a trade mission in Lusaka, Chiluba has argued unequivocally in recent regional meetings for Pretoria's membership in SADC and the Preferential Trade Area (PTA) of East and Central Africa. This enthusiasm has led some observers to liken Chiluba to Malawi's Banda.[18] In a perceptive analysis of the rapid changes in post-Kaunda Zambia, Stephen Chan notes that the new government is committed to revitalizing the domestic economy; in these efforts,

> it will seek, as a principal regional policy, closer economic contacts with South Africa. This will mean that, unlike the attempt made by SADCC to strengthen the horizontal or broadly based transnational economic linkages in the region, Zambia will seek to strengthen the vertical linkages between itself and South Africa, bypassing the emphasis on developing linkages with others in what has been the frontline. The image of the frontline will itself fade from the region, and the change of government in Zambia—with the new Zambian preoccupations—will hasten this process.[19]

Conclusion: Postapartheid and Beyond

The end of apartheid is unlikely to remove the basic power asymmetries and change the objective conditions of South Africa's neighbors. Graham

Evans goes so far as to argue that "despite the presumption in favor of equality with its neighbors, a new majority-ruled government . . . would be unable to resist the obvious benefits of being the key player in local balance-of-power politics. . . . In the absence of an external force the role of manipulating 'balances' seems preordained, whoever occupies the Union Building in Pretoria."[20]

Given this overall scenario, the future strategies by small states might still comprise a combination of collective and unilateral approaches, with the choice of each strategy determined by a convergence of domestic and international factors. With the disappearance of the primary source of regional adversity and the loss of external interest, the possibilities of regional security may lie in limited cooperative schemes based on prevailing economic structures and limited problem-solving security arrangements. As in the past, the dictates of minimalist approaches in the economic realm will far outweigh the temptations for grandiose and politically attractive maximalist mechanisms. The enduring tension between collective aspirations and individual reality is not peculiar to Southern Africa, but it promises to constitute the basis for regional foreign policies.

Writers who have posited that the region will fragment into economic zones reflecting the current areas of interdependence and functional cooperation proceed from the assumption that when national entities focus more on the problems of domestic insecurity, they have less impetus for a maximal approach to regional economic cooperation.[21] Subregional economic zones might be supplemented by cooperation around vital functional areas such as energy, water resources, and trade. Existing regional institutions with overlapping memberships such as SADC, the Southern African Customs Union (SACU), and the PTA could be restructured to fulfill more scaled-down and focused modes of economic cooperation.

Equally pertinent, since most of the existing institutions have often been competitive, restructuring them means that individual states would have to decide which ones are optimal for their needs without the pressures to conform to preestablished organizations. In the long run, such voluntary choices would resolve the awkward questions of overlapping memberships and duplication of efforts, which were the consequences, no doubt, of the past political efforts to straddle both sides of the racial chasm. The minimalist scenario is pessimistic about the long-term ability of SADC to move beyond what its predecessor achieved in the last decade; other than, perhaps, the continued resuscitation and expansion of the transport corridors in Angola and Mozambique, SADC's role in a postapartheid Southern Africa might be minimal. There will be no South African–led Southern African economic community because of the potential for conflicts over the costs and benefits between admittedly unequal economic partners.

This minimalist approach to regional cooperation goes against the grain of concerted efforts to fulfill SADC's EEC-style integration scheme. In spite of the logic of sustaining a credible cooperative arrangement to check South Africa's economic power, ill-conceived schemes requiring increased member obligations could obviate the practice of flexibility so crucial in past regional cooperation. Increasing skepticism about how diverse regional nations could come together under a treaty with binding economic and political obligations cannot be lightly dismissed even by well-intentioned policymakers. Furthermore, the competitiveness among regional states for bilateral concessions from South Africa and other external actors are telling lessons as to the futility of grand economic schemes for the future. Where bilateralism and strategic bargaining for unilateral advantages predominate, collective aspirations take a backseat.

The same concerns carry over to future regional security institutions. Some writers have envisaged for Southern Africa a future comprehensive security umbrella on the lines of the Conference on Security and Cooperation in Europe (CSCE). For example, Bernhard Weimer and Olaf Claus have contended that "the FLS alliance will disappear but its experience in consultation and conflict resolution will continue to prove valuable for . . . the possible formation of a Conference for Security and Cooperation in Southern Africa (CSCSA)."[22]

In spite of the need to improve channels of communication and consultation in the future, the CSCE framework with its elaborate structures defies anything these countries would contemplate. Like the economic arena, however, Southern Africa might need limited, manageable, and focused methods of conflict resolution rather than the comprehensive ones found in Europe. Thus, proposals for cooperation in arms control, mass migration and refugee issues, AIDS, and joint action on natural disasters appear to be feasible in light of prior practice of ad hoc and flexible political arrangements for problem solving. The experience of the FLS and SADCC, then, is how minimum coordination in issue areas can be fruitfully pursued within the realistic context of competing national agendas. In the postapartheid era, such limited arrangements could provide methods of regular consultation over specific issues as they also preempt the costs and conflicts associated with comprehensive security schemes.

Discussions of a comprehensive security arrangement for Southern Africa ought, more appropriately, to relate to the future and viability of the African continental system that finds institutional expression in the OAU. The white redoubt of South Africa was the "uncaptured" portion of the African interstate system awaiting future redemption.[23] A watertight security system for Southern Africa existing independent of the continental security system would, consequently, defeat the larger purpose of the liberation process. For despite its myriad flaws, the OAU has contributed to building enduring interstate norms, principally the respect for the independence of small states. In the African system, these norms have primarily laid the foundation that allowed the coalescence of states at the regional and functional level in their assorted attempts at economic integration. Reintegration of South Africa into the community of nations might therefore need to begin with the African interstate system rather than the more restrictive Southern African one.

Even without the economic benefits accruing from South African membership in the OAU, there certainly will be great value in its political participation; a "strength-in-numbers" argument suggests that a power with hegemonic proclivities would be better tamed within a larger organization than within a smaller one. Tied to continental interstate norms, a powerful South Africa is likely to have little incentive to play an overtly subimperial regional role. More critically, South Africa's role in the OAU might check the trend of its present leadership to fragment the OAU by promoting ostensible "regional powerhouses."

Notes

1. See, for example, Douglas Anglin, "Southern Africa Under Siege: Options for the Frontline States," *Journal of Modern African Studies* 26, no. 4 (1988), pp. 549–565; Leslie H. Brown, "Regional Collaboration in Resolving Third World Conflict," *Survival* 28, no. 3 (1986), pp. 208–220; and Mahnaz Z. Ispahani, "Alone Together: Regional Security Arrangements in Southern Africa and the Arabian Gulf," *International Security* 8, no. 4 (1984), pp. 152–175.

2. Sam Nolutshungu, "Strategy and Power: South Africa and Its Neighbors," in Shaun Johnson, ed., *South Africa: No Turning Back* (Bloomington and Indianapolis: Indiana University Press, 1989), pp. 335–352.

3. I. William Zartman, *Ripe for Resolution: Conflict and Intervention in Africa*, updated edition (New York: Oxford University Press, 1989), p. 179.

4. John Saul, "From Thaw to Flood: The End of the Cold War in Southern Africa," *Review of African Political Economy* 50, no. 2 (April 1991), p. 157.

5. Quoted in the *Plain Dealer,* November 3, 1991.

6. Thomas Callaghy, "Africa and the World Economy: Caught Between a Rock and a Hard Place," in John Harbeson and Donald Rothchild, eds., *Africa and World Politics* (Boulder: Westview Press, 1991), pp. 39–68.

7. Christopher Coker, "'Experiencing' Southern Africa in the Twenty-first Century," *International Affairs* 67, no. 2 (1991), p. 285.

8. Peter Vale, "The Search for Southern Africa's Security," *International Affairs* 67, no. 4 (1991), p. 703.

9. Simba Makoni, "Interview," *Africa Report* 35, no. 3 (1990), p. 35.

10. Quoted in the *Daily Nation,* June 10, 1990.

11. Quoted in *Foreign Broadcast Information Service (FBIS)–Africa*, September 10, 1990.

12. Douglas Anglin, "Southern African Responses to Eastern European Developments," *Journal of Modern African Studies* 28, no. 3, p. 438.

13. Colin Stoneman and Carol Thompson, "Southern Africa After Apartheid: Economic Repercussions of a Free South Africa," *Africa Recovery Briefing Paper* No. 4, December 1991, p. 10.

14. Through a wide range of statements, especially at regional meetings, the ANC has assured SADC states that it abjures past heavy-handed practices. The issue that pessimists dwell on, however, is whether such promises will be translated into actual policies when the ANC is confronted with both the realities of power and the temptations of hegemony. As Peter Vale argues in Chapter 10, the result will depend in large part on whether South African diplomacy is transformed and a new image of South African diplomacy is forged.

15. For a compilation of such contacts, see Colleen L. Morna, "South Africa: The Pariah's New Pals," *Africa Report* 36, no. 2 (1991), pp. 28–30.

16. Quoted in the *Washington Post*, June 9, 1991. See similar comments by de Klerk in the *New York Times*, October 27, 1991, and in *Africa Confidential*, June 19, 1992.

17. Quoted by the BBC, November 25, 1991.

18. See, for example, Melinda Ham, "Zambia: End of Honeymoon," *Africa Report* 37, no. 2 (1992), pp. 61–63; and Steve Kibble, "Zambia: Problems for the MMD," *Review of African Political Economy* 53, no. 2 (1992), pp. 104–108.

19. Stephen Chan, "Democracy in Southern Africa: The 1990 Elections in Zimbabwe and the 1991 Elections in Zambia," *Round Table*, April 1992, p. 199.

20. Graham Evans, "Myths and Realities in South Africa's Foreign Policy," *International Affairs* 67, no. 4, p. 716.

21. Tony Hawkins, "Hard-headed Realism," *Southern African Economist* 3, no. 5 (1990), pp. 7–9; Gavin Maasdorp, "A Changing Regional Role for SADCC?" *Harvard International Review* 12 (Fall 1989), pp. 10–13. For a general explanation of the tensions between nation-building and regional cooperation, see Stephen John Stedman, "Conflict and Conflict Resolution in Africa," in Francis M. Deng and I. William Zartman, eds., *Conflict Resolution in Africa* (Washington, D.C.: Brookings Institution, 1991), pp. 377–378.

22. Bernhard Weimer and Olaf Claus, "A Changing Southern Africa: What Role for Botswana?" in Stephen John Stedman, ed., *Botswana: The Political Economy of Democratic Development* (Boulder: Lynne Rienner Publishers, 1993), pp. 185–202.

23. I. William Zartman, "Africa as a Subordinate State System in International Relations," *International Organization* 21, no. 3, pp. 545–564.

12

South Africa:
The Political Economy
of Growth and Democracy

Robert M. Price

*A society divided between a large impoverished mass and a small
favored elite results either in oligarchy (dictatorial rule of the
small upper stratum) or in tyranny (popular-based dictatorship).*
—Seymour Martin Lipset[1]

By the terms of conventional wisdom the prognosis for democracy in
South Africa is decidedly unfavorable. If extremes of wealth and poverty
create infertile soil for democracy to take root, as is commonly asserted,
then over eighty years of white rule would seem to have bequeathed to
South Africa a particularly inhospitable environment to sustain a democratic order. Its poverty profile may well reflect the most extreme combination of affluence and impoverishment of any country in the contemporary world—and one that is especially explosive politically because it
coincides with a racial, and thus highly visible, divide. White wealth and
black poverty in South Africa produces a Gini coefficient (the standard
measure of income inequality) that is the highest of any of the fifty-seven
countries in the world for which data are available.[2] Sixty percent of all
black South Africans (80 percent of those living in homelands) have incomes below the minimal living level (MLL), according to a study by the
Carnegie Commission.[3] The basic contour of South Africa's poverty profile—the huge gap between white wealth and black poverty—was somewhat eroded during the 1970s but has been reinforced in the past decade.

A stagnant economy for much of the last ten years, resulting in large part from a combination of domestic political upheaval and internationally imposed sanctions, has produced a decline in personal income and soaring rates of black unemployment.

The prognosis for economic growth in a country with a poverty profile like South Africa's is no better than for democracy. Majority impoverishment breeds an atmosphere of populism that is hostile to the adoption of growth-producing economic policies. Where mass poverty breeds desperation, the time horizon of the majority is not compatible with economic policies that would sustain growth over the long run. When the poor are politically mobilized, pressures are created for economic policies geared to immediate consumption and expenditure rather than investment and production. The populist politician sustains a hold on power by distributing society's material resources in the short run while bankrupting it in the long run. The history of postcolonial sub-Saharan Africa is a textbook portrait of this form of political economy.

Despite the apparent inhospitable nature of South Africa's socioeconomic environment for either democracy or growth, the thesis of this chapter is that in fact there is much in the country's contemporary situation that can provide a foundation for both. While the obstacles are substantial, in comparative terms South Africa ought to be seen as offering perhaps the best chance for success of the current cases of transition. Relative to the countries of Eastern Europe, the former Soviet Union, and the states of tropical Africa, South Africa can be seen as possessing the most substantial social base for democracy and economic base for growth.

South Africa's "advantages," which I will turn to presently, can certainly be undermined by the political consequences of its poverty profile, but that is not inevitable. I will argue that the key to sustained democracy and growth in the face of inevitable populism will be found in the arena of "governance." A postapartheid governing elite will need to manage populist pressures so as to simultaneously build legitimacy for democratic institutions and protect investment, production, and thus growth-generating economic activities. This will be *the* challenge of governance in South Africa once the demise of white rule is complete. The ability of a new government to meet the challenge will be decisively affected by the institutional arrangements that emerge out of negotiations. The matter is not reduced to whether these institutions meet a democratic standard. Some "democratic" institutional arrangements will facilitate meeting the challenge of governance while others may not. The latter, although formally democratic, may, because they do not meet the challenge of governance, contribute over the medium term to democracy's demise.

The Social Bases of Democracy

South Africa possesses two important social assets for the establishment of a democracy: the acceptance of South Africa as a single political community by the dominant political forces in society, and the existence of a vibrant civil society. The first means that it is unlikely to experience in the postapartheid era the type of corrosive politics of ethnicity or "tribalism" that is today witnessed in the former Yugoslavia, certain republics of the former Soviet Union, and in most of the postcolonial states of sub-Saharan Africa. The second is important for two reasons: it provides a social base for maintaining electoral competition, the essence of representative democracy; and it contains a mechanism for ameliorating the populist pressures that otherwise will threaten both the legitimacy of a new democratic order and the prospects for implementing growth-generating economic policies.

Political Cleavage and Political Community in a New South Africa

It is becoming commonplace to describe South Africa as a "deeply divided" society,[4] one in which intra-African ethnic cleavages require special constitutional arrangements to prevent a new democratic order from being destroyed by the centrifugal and violent forces of communalism. Not just the white ruling party, which for over a decade has proclaimed South Africa to be without a genuine majority because it is a country of cultural minorities, promotes this view, but respected academics at home and abroad, as well.[5] Constitutional engineering—consociational schemes embodying minority vetoes over government policy, federalism or regionalism with a strong devolution of power, electoral systems that allow small party parliamentary representation and force coalition governments—is said to be necessary to avoid a situation in which fears of "domination" will splinter South Africa, probably violently, along the lines of its ethnocultural pluralism.

Although the image of South Africa as a society "deeply divided" along lines of intra-African cultural pluralism increasingly goes unquestioned, there is actually little or no direct evidence to support this picture of the country's political cleavages. The few attitude surveys that make an effort to tap the relationship between culture and politics in South Africa indicate a rejection by black South Africans of ethnicity as a basis for political membership and identity.[6] For example, in one study of high school students in Soweto, respondents were asked what name they would prefer to describe "their people." Not a single one of the randomly selected students interviewed offered an ethnic label (64 percent preferred the term

"African" and an additional 13 percent "Blacks").[7] Nearly 90 percent of the sample said that black South Africans should "form one nation irrespective of tribal origin," and nearly 70 percent answered no when asked if tribal identity should be preserved.[8] Similar results were obtained in a recent nationwide study of black trade union shop stewards, a group of over 20,000 factory-floor-worker activists. In interviews conducted during 1991, it was found that 86 percent identified themselves mainly as South Africans and only 5 percent mainly as members of an ethnic group.[9]

More significant, with respect to black politics there is, with one exception, not a single political, or even quasi-political, organization with any real following that defines its membership, mobilizes its followers, or shapes its demands in ethnic terms.[10] In other words, the existential landscape of South African black politics is simply not contoured by ethnic cleavage. Political organization and mobilization, the things that count in the world of competition and conflict, do not reflect a terrain scarred by multiple and deep politico-cultural fissures, as both Pretoria and various liberal academics would have us believe. The one exception is, of course, the Inkatha Freedom Party of Chief Mangosuthu Buthelezi and the KwaZulu government. But is the existence of Inkatha, and of the political reaction to it among black South Africans, evidence of the fracturing of African political allegiance along ethnic lines? I think not. First, there is the question of the extent of Inkatha's following. Virtually every survey of political allegiance in the last two years places Inkatha support at less than 10 percent of African adults countrywide. A survey involving interviews with urban, rural, and homeland residents, carried out in September–October 1991, found that only 3 percent of the African population would vote for either Inkatha or its leader, Chief Buthelezi (67 percent said they would vote ANC).[11] More recently, an interview survey conducted in May 1992, in all major urban centers (including Durban), revealed again that only 3 percent of adult urban Africans would vote for Inkatha.[12] Even more striking, 71 percent of the randomly selected sample indicated that they rejected Inkatha "completely and on principle." The implications of these surveys is that Zulu-speaking South Africans living outside of Natal, and even those within Natal's urban areas, reject Inkatha and its ethnically chauvinist appeals. This is consistent with various estimates that Inkatha support is relegated largely to rural Natal, i.e., the area that falls under the sway of the KwaZulu patronage machine and police force. For example, a random sample survey conducted in 1985 in Durban's largely Zulu-speaking townships found support for Inkatha and Buthelezi among only 4.8 percent of the population (in contrast to 54.2 percent for Mandela).[13]

The reaction of the non-Zulu African population to Inkatha's blatantly chauvinist rhetoric also indicates the weakness of tendencies toward ethnic fragmentation in contemporary South Africa. Have other Africans reacted to this use of Zulu nationalism by seeking political protection in their own ethnicity? On the contrary, there is simply no evidence that Inkatha's efforts at ethnic mobilization have produced a counter-Zulu ethnic mobilization, fracturing the African population into new political organizations that make exclusivist ethnic appeals to Xhosa, Tswana, Venda, Pedi, Sotho, and the like. Rather, the multi- or, as it prefers, nonethnic ANC has held its dominant position among the African population. In the 1992 survey that found only 3 percent urban African support for Inkatha, the ANC received the support of 70 percent (only 3 percent rejected it "completely and on principle"). In sum, neither the level of support for Inkatha among Zulus, nor the political reaction by other ethnic groups to the mobilization of Zulu chauvinism, suggests that politics in a "new" South Africa will be defined along lines of intra-African ethnic cleavage.

Proponents of the thesis that South Africa is a deeply divided society are not deterred by the lack of evident political cleavage along ethnic lines. Their arguments are based primarily on a projection onto South Africa of the political experience of Africa's other postcolonial states. They assume that once the "glue" of white supremacy has been removed, the black population will fragment politically along ethnic/tribal lines, as occurred in places like Nigeria and Uganda. Thus, according to Donald Horowitz, "Politics all over Africa . . . has a strong ethnic component. . . . What is true of Zimbabwe, Nigeria, Zambia, Kenya, and Mauritania is also likely to be true of South Africa."[14] "Eliminate white domination, and intra-African differences will be particularly important."[15] In the same vein, Arend Lijphart asserts that "there is no doubt" that black South Africa is "an ethnically plural society on a par with most of the black states in Africa." While "these ethnic cleavages are currently muted by the feelings of black solidarity in opposition to white minority rule," Lijphart continues, "they are bound to reassert themselves in a situation of universal suffrage and free electoral competition."[16] Such simple and unexamined projections onto South Africa of political life elsewhere in sub-Saharan Africa fail to take into account the radically different experience of South Africa in the nineteenth and twentieth centuries—an experience that makes analogies with other countries in Africa dangerously misleading.

In the south, the nineteenth century was characterized by a continuing series of wars of dispossession. These resulted not only in a loss of sovereignty for the traditional African political communities of the South African interior, but also in the wholesale displacement of communities from their indigenous habitats, and in extensive land alienation. The traditional social systems were left with inadequate land to support their populations. This had dire consequences for the functioning and solidarity of traditionally defined political communities. The material basis for kinship and village solidarity, upon which traditional African social systems are built, lay in the system of "usufruct" land tenure. By virtue of membership in a kin group, a person's access to communally "owned" farming land was assured. The alienation of most African land by the end of the nineteenth century vitiated this kinship-based guarantee, and thus fundamentally undermined the traditional system of land tenure and with it the entire traditional sociopolitical system. In much of Africa to the north of South Africa, colonial imposition, while involving a loss of sovereignty, left the traditional systems otherwise intact. These, with their core system of communal land tenure, the centrality of kinship organization, and the overriding identification with village communities, continued to function, providing valued resources to their members throughout the colonial and into the postcolonial eras. The very different historical experience of South Africa compared to its neighbors to the north, and the resultant difference in their contemporary social organization, does not support an expectation of similarity with respect to the nature of political identification in each.

In the twentieth century, the differences between South Africa and other territories in sub-Saharan Africa were reinforced. Africans in South Africa were, beginning in the last decades of the nineteenth century, drawn and forced into a process of proletarianization and urbanization. African society, in very significant proportion, became part of a modern industrial order. The same cannot be said for the rest of Africa, whose societies emerged out of colonialism as still essentially peasant. Some thirty years after the era of independence, South Africa remains the only significantly industrialized country in sub-Saharan Africa. There is no reasonable basis to assume that the tribal politics of Africa's peasant societies provides a benchmark for expectations about the future shape of politics in industrial South Africa.

Finally, the apartheid experience of the last forty years can be expected to shape the way ethnicity is politicized in South Africa. Simply put, by trying to use traditional identities and ethnicity as a means to divide and thus better control South Africa's black majority, the white minority state has delegitimized traditional ethnicity as a basis for political organization and mobilization. With the exception of the "homeland" collaborationist elite, politically aware black South Africans have, for four decades, reacted to Pretoria's separate development designs by explicitly and fervently rejecting politicized ethnicity. This is reflected today in the total absence of ethnically organized political groups, save those that are tied to homeland governments. Those who expect South African politics to fragment along traditional ethnic lines, once the demise of white political supremacy is complete, ignore or dismiss, without the slightest rationale or evidence, the impact of nearly forty years of intense political socialization in which a South African political identity has been held up as appropriate and juxtaposed to an illegitimate, and reactionary, ethnicity.

The projection of an image of impending ethnic strife on the South African political landscape is not merely analytically misleading but also potentially consequential for shaping the future of South Africa. Those who propound the "African ethnicity thesis" nearly always combine it with proposals for constitutional engineering to avoid the centrifugal political forces and intense conflict that are associated with ethnic-communal politics. The particulars of these proposals differ, but they all have one element in common: they seek to avoid fears of domination, perceived as the engine of ethnic strife, by creating structures that will ensure ethnic groups, as groups, a share of political representation, power, and public resources. Such efforts may well serve a useful purpose under conditions of ethnic fragmentation and strife, but they are counterproductive in situations like South Africa's, which do not exhibit the ethnicization of political competition. By structuring the political game in a way that provides significant resources to political groups organized along ethnic lines, such efforts have the effect of promoting ethnic "entrepreneurs," and providing incentives for ethnic organization and mobilization. The constitutional schemes to prevent ethnic strife in South Africa, if adopted, are thus likely to bring into being exactly the politicized ethnicity that their proponents supposedly wish to avoid. In sum, the constitutional proposals put forward by those who claim South Africa is a society deeply divided by ethnic cleavages should be viewed as a threat to a stable and democratic future.

Civil Society and Democracy in South Africa

It is generally accepted that sustained democracy requires something that is termed "civil society." At its simplest, what is involved here is the notion that more is required for a functioning representative democracy than the holding of regular elections for the selection of officeholders. These are necessary elements, but they contribute little to the substance of democracy unless the organization of political interests takes place independently from the state—that is, outside the control of officialdom. If governing groups can use their power to form, shape, and control the organization of political interests in society, they destroy the popular sovereignty that the election of officials is supposed to bring into being. It follows, then, that democracy requires a realm of "social space" in which people can act autonomously from the state in their efforts to coalesce, aggregate, organize, and push their perceived political agendas. This realm of autonomy or immunity from state control is what is meant by the term "civil society."

State officials everywhere have an interest in using the awesome power that their offices provide—coercive, material, symbolic, and informational power—to shape the organization of political interest in a manner that will maintain their own incumbency. In other words, state officials are inherently in conflict with civil society; their natural proclivity is to vitiate the realm of autonomous political action. Hence, only when there exist politically oriented social organizations that are consciously jealous of their autonomy from the state, and that possess the resources to protect that autonomy, will a political system possess a civil society. Only then will the realm of political autonomy from the state, so necessary for a functioning representative democracy, be protected from the omnivorous proclivities of officialdom.

An extraordinary thing about contemporary South Africa is that it is one of the few countries currently undergoing transition away from authoritarianism that actually possesses a vibrant civil society. That civil society was born in the fight against apartheid that marked the decade of the 1980s. The continuous political struggles of the period were driven forward by a concurrent process of social change whose hallmark was organizational effervescence within the black community. The period witnessed a veritable explosion in associational life. It gave birth to new organizations of every variety—community, youth, women's, labor, student, political— which by mid-decade honeycombed the social fabric of all but the smallest and most remote of the country's black townships. The new associations, the most politically significant of which are the trade unions and township civic associations, articulated a sense of mounting grievance, mobilized collective actions against established authority (rallies, demonstrations, strikes, boycotts), and offered a new basis for governance within the urban township communities. By 1984, the array of new community-based grassroots organizations had become constituent elements in a full-scale insurrection against white political supremacy. Ultimately, the political pressure these new associations were able to bring to bear, in combination with the international economic sanctions their actions stimulated, forced the state to agree to negotiate an end to white minority rule.

A variety of factors suggest that the associations that played such a central role in the political upheavals of the 1980s will emerge out of the period of political transition as the core of a new South African civil society.[17] From the vantage point of the emergence of civil society, the most significant thing about these new associations is that they were born both in opposition to the state *and* independently of the African National Congress. Although by 1985 virtually all of the grassroots community-based organizations were closely aligned with the ANC-led political struggle against white supremacy, they were formed independently of the ANC and developed outside of the ANC's formal political and bureaucratic apparatus.

The formative experience of South Africa's new organs of civil society, in particular the trade unions and civic associations, has been key to their ability to maintain an independent organizational existence. Their leadership cadre derived its strength and legitimacy from the factory floor and the community grassroots, not from the apparatus of a political movement. The roles that grassroots organizations and their leaders played in the struggle for liberation was not only important but also self-directed. And, a decade-long intense political struggle has bequeathed to the unions and civics, as well as to student, women's, youth, and other community-based groups, a deep multilayered leadership cadre, operating at national, regional, local (township), and even neighborhood levels. The leadership is also drawn from the full range of layers of social stratification. The new civil society in South Africa is not the preserve of a thin stratum of middle-class intelligentsia, as appears to be the case for much of Eastern Europe, the former USSR, and tropical Africa.

The breadth and depth of the black leadership cadre has had two significant consequences. In the period from 1986 to 1990, it undermined the efforts of the state to eliminate the organizations of an incipient civil society. Leadership ability and experience had spread so widely and deeply within the black community that obliterating a narrow elite stratum was no longer sufficient for defeating the state's political opponents. In the period after 1990, when the now legalized ANC with its already extensive exile apparatus returned from abroad, it was virtually impossible for it to absorb the entire local leadership cadre that had arisen during the 1980s. Ironically, the considerable difficulty the ANC experienced in integrating local leadership into its exile-dominated organization, while it may have temporarily weakened the liberation forces, strengthened the chances for democracy by facilitating the autonomy of an incipient civil society. In sum, the community-based organizational efforts had grown too large for them to be either easily repressed by the state or absorbed by the dominant political movement.

Although the particular historical circumstances that contributed to the emergence of grassroots associations and to the maintenance of their organizational integrity and independence are the key elements in the creation of an incipient South African civil society, there are several features of the South African situation that will likely help sustain it. A supportive mass media is one. The decade of the 1980s witnessed the birth of a vibrant "alternative," and at the same time professional, press. By giving extensive coverage to developments at the grassroots level in the black community, newspapers like the *Weekly Mail* and the *New Nation* have made both the existence and the autonomy of community-based associations part of the public consciousness. While politically sympathetic to the ANC, the "alternative press" are not party papers, and their reporting has been very sensitive to threats to the independence of community-based organizations. In that way they have served to generate and sustain an environment that encourages the continued autonomy of civil society.

The prospects for maintaining a realm of civil society are also enhanced in South Africa by the existence of an array of supportive, influential individuals and institutions. Most significant in the latter category are the Institute for a Democratic Alternative in South Africa, founded in 1986, and the Institute for Multi-Party Democracy, established in 1991. Both organizations seek to foster a political culture supportive of democracy through a vigorous program of public education and research. At their numerous conferences, symposia, workshops, and other forums of public education, the exploration of the civil society phenomenon and its importance for democracy is a prominent theme.[18] South Africa's extensive academic and quasi-academic community has also been an important resource in building and strengthening the organizational capability of grassroots associations. In the past decade and a half, "progressive" intellectuals and academics have provided staffing, technical assistance, and general ideological support for community-based organizations. In the early 1980s, they were deeply involved in the new independent trade union movement. More recently, an infrastructure of private but nonprofit policy and development firms focusing on urban issues has emerged. As civics move from resistance to reconstruction, organizations like the Centre for Policy Studies, PlanAct, and CORPLAN will be available to supply technical assistance in the areas of town planning, housing construction, and local-level negotiations.

Another element strengthening organizations of civil society in South Africa is their access to funding, much of it from foreign sources. In the current era, genuine community-based grassroots organizations are looked upon with considerable favor by philanthropic foundations, labor unions, governments in the industrialized countries, and international agencies. The level of resource transfers from these sources to South Africa will increase dramatically when and if there is a democratic political settlement. At that point, donor agencies will seek ways to dispense resources in a politically neutral manner. The formal nonpartisan stance of community organizations like civics will thus place them in an excellent strategic position to be the recipients of major new resource flows.

External support from individuals and groups, both foreign and domestic, helps strengthen community-based organizations, but the most important factor sustaining South Africa's civil society is of a different type. The organizations of civil society, especially the trade unions and the civics, possess important political resources that the national political organizations lack. The political movements may have broad popular support, but they are lacking in grassroots organization.[19] That is precisely what the civics and unions have, and it can be used both to sustain their autonomy from the state and to influence its policies. In the future, when (if) electoral politics takes hold, the behavior of voters will likely be mediated through these new organizations of civil society. Political parties will vie for the support and endorsement of the local grassroots associations. Attending to the issues raised by them will constitute the surest route to electoral success. As long as they remain autonomous from the ruling political party, as they have so far in South Africa, they will provide the social base for the maintenance of electoral competition.

Prospects for Economic Growth

Students of South Africa, their vision dominated by ten years of economic stagnation, an eroding stock of fixed capital, an unemployment rate that could hit 50 percent by century's end, persistent annual inflation of 15 percent, on top of apartheid's legacy of poverty and inequality, have been overtaken by pessimism about South Africa's growth prospects after the transition from white rule. But, while the country's economic problems are real, observers of South Africa might find in a comparative perspective a helpful antidote for their gloom. In the countries of Eastern Europe and the former Soviet Union, virtually every institution of modern market production and exchange must be created from scratch and, to be successful, probably all at once: property rights, financial institutions, a credit system, capital markets, physical infrastructure, taxation instruments, a private savings system, a convertible currency, and basic cultural orientations like a work ethic and the willingness to take initiative. In this light the obstacles to South Africa's economic growth appear minor and the prospects for success, if not bright, at least reasonable. Unlike its counterparts in the former communist world, South Africa possesses a fully functioning market-based industrial economy, albeit one that is currently strained and requiring major structural adjustments. But the point is that for South Africa the challenge is making adjustments, not, as in the former communist countries, the need to create in just a few years what in other societies took centuries.

In addition to the full array of market institutions, and corresponding state mechanisms for fiscal and monetary policy, South Africa will be able to draw on a number of distinctive assets in pursuit of postapartheid economic growth. These include (1) a major share of the world's reserves in nearly all important industrial minerals; (2) a highly developed and modern transportation and communications infrastructure; (3) a significant pool of scientific, engineering, and professional manpower; (4) a very favorable credit situation, the ironic result of international financial sanctions that after 1985 forced the payoff of medium-term foreign debt while blocking access to new credit; and (5) in the medium term, a natural and increasingly sizable market for its manufactured goods in the Southern African region.

One of the more positive aspects of the South African situation as far as growth potential is concerned is the economic pragmatism that has been displayed by the leadership of the political movement likely to dominate after the first postapartheid election. The ANC has been repeatedly and heatedly criticized by the South African business community and media for its supposed commitment to the "outdated" notions of nationalization and socialism. Indeed, the ANC has been unable to completely cut its tethers to these notions annunciated in the 1955 Freedom Charter; in the contemporary context they are powerful symbolic code words for much of its popular constituency, meaning a commitment to social justice and economic redistribution. Yet in its economic policy discussions—in numerous workshops, conferences, publications, speeches, and interviews—the organization has for the past three years revealed considerable flexibility and realism with respect to economic models and future policy.

Central to the ANC leadership's emergent economic vision is the notion that improvement in the living conditions of the black majority (termed redistribution in the South African debate) can occur only through economic growth.[20] With growth viewed as the foundation for redistribution, a set of related premises has also been accepted: that jump-starting the growth engine will require attracting foreign capital and mobilizing domestic saving, which in turn requires accepting that businesses must be permitted a competitive return on their investments; that the private sector will play a major role; and that the rights of private property must be respected if a positive investment climate is to be maintained. Nationalization has been relegated to the status of simply one among many potential instruments for job creation and controlling the allocation of economic surplus. But the real trend in ANC economic discussions is toward industrial policy and "codetermination," the kind of social pact that has characterized relations between business, labor, and the state in Germany and through which trade-offs between wages, productivity, and social welfare spending are agreed upon.

Some elements of the South African business community and many of the liberal media have greeted the ANC's refusal to reject nationalization completely and on principle, as well as its preference for a strong economic role for the state, with howls of horror, as if its economic vision was hatched in the old GOSPLAN buildings. But the ANC is modeling the future not on the old communist command economies but rather on what Lester Thurow calls the "communitarian capitalism" of Germany and Japan, in which the state plays a significant "guide role" in the economy.[21] Ironically, while the ANC looks for guidance to the robust capitalist economies of the 1980s, the South African government, business, and media insist that the only rational way to go is pure laissez-faire, an approach practiced only imperfectly by the declining capitalisms of Britain and the United States.

The real threat to South Africa's prospects for economic growth lies not in the economic but in the political arena. It is the combination of poverty, inequality, and political mobilization, noted at the outset of this chapter, that mounts the biggest challenge to South Africa's future. As a leading ANC economist noted, "The resources that will be released [through a 'postapartheid dividend'] will hardly be enough to redress poverty on the scale necessary to overcome the crisis [black deprivation]."[22] As a consequence, will the first postapartheid government, presumably one in which black South Africans predominate, be able to resist the political pressure to seek immediate redistribution through policies of job creation, social spending, protectionism, and confiscatory taxation, while ignoring inflation, waste, and inefficiency? If it cannot, it is certain that the flow of foreign and domestic investment required for growth will not materialize. Some short-run and marginal improvement in black living conditions might be obtained, but in the longer term, continued economic decline and fewer resources for poverty reduction would result.

The populist challenge extends beyond the arena of economic rationality and growth. It will also pose a threat to political stability and to the sustainability of a democratic order. For unless the first postapartheid government can deliver on at least some black expectations, it is unlikely that the current ANC leadership, with its commitment to economic rationality and a market-oriented and open economy, will be able to survive an attack "from the left." Both within the organization and outside it there exist potential challengers who are far more attracted to an economic perspective that emphasizes autarkic "self-reliance" policies, and that has much less tolerance for the rights of private property and the acceptance of the "realities" of the market. Even if the ANC leadership could hold off its populist challengers, the threat to democracy would probably be devastating. For among South Africa's majority the legitimacy of a new democratic order will have to be earned. A new constitutional order must demonstrate that it can respond effectively to the crisis of poverty and inequality, or it is likely to be swept aside by the political force of a populism driven by fierce social anger.

The Postapartheid Governance Challenge

South Africa's postapartheid political economy, shaped by a combination of the country's severe poverty profile and high levels of expectation and politicization, defines certain clear parameters of action for the first postapartheid government. It must, at one and the same time, adopt policies that aggressively address at least some of the material expectations of the black majority, while not allowing those policies to undermine the fundamental investment orientation of economic policy. There is, I believe, room for these contradictory imperatives to be effectively dealt with, although it will not be easy.

In the last decade and a half, black deprivation has taken on specific political shape in three areas in particular: housing, education, and wages. Grievances in each area gave powerful impetus to the antiapartheid struggle during the 1980s, and each gave rise to forms of organized resistance that today are core components of South Africa's fledgling civil society. Inadequate township housing provided the context for the development of the civics movement, demands for the elimination of Bantu Education produced a student movement, and apartheid's low-wage system was the foundation for the trade union movement. Housing, education, and wage demands could in a postapartheid context once again be the basis of anti-state mobilization. They could also, however, provide the basis for a postapartheid government to consolidate a new democratic order and dampen the populist threat to economic growth.

Massive programs of urban housing development, school construction, teacher training, and adult literacy, combined with a "social pact" between labor and business entered into through a codetermination process, offer the best hope for combining short-term improvements for the black population with labor discipline, wage restraint, and increased productivity. The emphasis in housing and school construction should be on user-built projects, in which community organizations would play a crucial role. The cost of building materials would have to be covered largely by foreign donors, although a special national real estate levy on existing residential and commercial property could be an additional source of funds.[23] The depressed South African construction industry can produce much of what is needed in the way of building materials, and donated technical skills in architecture, design, and building trades can be mobilized from the white sector of society. With unemployment in the urban townships at over 40 percent, it becomes feasible for labor costs to be donated by community members themselves. And most important of all, the strong civic association movement provides the organizational resources, on the ground, to mobilize and direct the local residents in the building of their own homes and schools.

The political benefits of such community-built housing schemes are multiple. Housing is not only a high priority for the black community, but it is the kind of benefit that provides tangible evidence that some measure of relief is being delivered to the previously deprived. While not everyone can obtain a new house simultaneously, massive ongoing construction activities will provide credibility to the promise of one in the near future. User-built houses and community-constructed schools also create the opportunity to impart valuable skills in the construction and building trades to previously unskilled residents of urban townships. The demand for building materials will rejuvenate a depressed construction industry. The mobilization of white professionals to assist black communities in reconstruction can play a supportive role in the all-important task of racial reconciliation. A national adult literacy campaign can provide a role for student organizations akin to the role of civics in the area of housing. It is a low-cost means of mobilizing thousands of young political activists in the task of social reconstruction. And, like the housing schemes, it can provide dramatic and visible evidence that a new government is delivering.

The most immediate focus of populist pressure will likely be in the arena of labor relations. The organized sector of the working class holds a strategic position in the economy, has a history of militancy, and has demonstrated its capacity to disrupt the production process on a large scale. The best hope for avoiding frequent strike actions and keeping wages and productivity in balance is the creation of institutions that draw both labor and management together to bargain over the major interrelated issues of modern political economy. In such arrangements of codetermination, the state participates as arbiter and as guarantor of agreements reached. Interest groups organized nationally interact under government auspices to strike "peak bargains" that involve social, fiscal, monetary, and incomes policy. Those industrial countries that practice this form of "democratic corporatism" have the best records with respect to combining economic growth with social welfare.[24] South Africa already possesses the key national "peak associations" necessary for instituting codetermination, in the form of large labor federations, especially COSATU and the several national business chambers in mining, industry, and commerce. The leadership of COSATU is already thinking in these terms, as witnessed by its support of a National Economic Forum to set economic policy after apartheid.[25]

Although South Africa possesses the grassroots organizations that will permit a new government to aggressively address the "crisis of deprivation," the form of government that emerges out of negotiations may well be not equal to the task. The kind of effort required will demand a strong and coherent central state, one that can make decisions, mobilize resources, and galvanize and direct energies toward nationally coherent tasks. Students of successful democratic corporatism and codetermination in Europe emphasize that one core requirement is a centralized government.[26] Assuming that a democratic constitution emerges out of negotiations, will it embody an institutional arrangement capable of governance of this type? This is far from clear. Pretoria, the Democratic Party, and a number of liberal academics have been pushing a variety of arrangements, all of which have one thing in common. They encourage weakness at the national center—either through a radical devolution of power to peripheral units (federalism/regionalism), or through legislative and/or executive immobilism (vetoes, collective executives, proportional representation, and multiparty coalition cabinets).

For the liberal academics, these preferences stem from misguided notions about the implications of South Africa's cultural pluralism. Misjudging the nature of political differences, the liberals focus exclusively on the need to avoid what they assume is imminent ethnic strife, and in the process they design constitutional solutions that completely ignore the tasks of redistribution and governance. Fearing the consequences for white living standards of majority rule, Pretoria and the National Party seek to bequeath to South Africa a central government too weak to intervene socially and economically. In either case, state weakness will spell the inability to deliver some measure of relief to the black population in the short term. The likely consequence will be to politically undermine the most rational and pragmatic portion of the ANC leadership and, more significantly, to discredit the democratic alternative for South Africa. A form of democracy that lacks the potential for effective governance, relevant to South African conditions, will be unable to develop the legitimacy necessary to sustain itself. It will likely be swamped by the highly mobilized and politicized masses whose expectations it has dashed.

Analogies figure prominently in discussions on South Africa's future. The South African government as well as many liberal academics frequently makes reference to the experiences of Nigeria, Uganda, and other sub-Saharan African countries as "models" that need to be avoided in engineering a South African constitution. They might be better served, however, by looking at the German experience during the interwar years. The Weimar Republic was created with one of the most democratic constitutions of its day. It lasted less than five years. The particular institutional arrangements of Weimar democracy—a dual executive, proportional representation, multiparty coalition cabinets—produced a weak central state. The republic's government was consequently incapable of decisive action in dealing with the socioeconomic crisis that gripped Germany in the years following World War I. The result was to discredit democracy in the eyes of the German people and to pave the way for National Socialist dictatorship.

Notes

1. *Political Man* (New York: Doubleday, 1959), p. 50.
2. This result is for 1978. More recent calculations by the IMF suggest that a degree of greater equality was achieved in the early part of the 1980s. Yet the country's Gini coefficient remained higher than that of any of the developed Western economies, and only the Latin American cases that suffer highly uneven income distribution came close to the South African coefficient. See International

Monetary Fund, *Economic Policies for a New South Africa*, Occasional Paper No. 91 (Washington, D.C., January 1992), p. 4.

3. The Carnegie calculation was done on data from 1980. A study using more recent data reports that in 1989, 40 percent of South Africa's total population lived below the MLL. This, however, underestimates the overall and black poverty level because the data upon which these calculations were based excluded the four nominally independent homelands. See Francis Wilson and Mamphela Ramphele, *Uprooting Poverty—The South African Challenge: Report for the Second Carnegie Inquiry into Poverty and Development in Southern Africa* (New York: Norton, 1989); and IMF, *Economic Policies for a New South Africa*, p. 5.

4. Introduced by Arend Lijphart in his study of consociationalism in the Netherlands, the concept "deeply divided" society refers to systems comprised of essentially self-contained (unintegrated) cultural/ethnic or religious communities. Lijphart has subsequently applied the concept to South Africa in characterizing the cultural pluralism of the country's black population. Lijphart uses the terms "deeply divided," "segmented," and "plural" society interchangeably. Several South African analysts, especially Hermann Giliomee, have adopted Lijphart's usage in their discussions of appropriate constitutional solutions for dealing with their country's problems. See Arend Lijphart, *The Politics of Accommodation: Pluralism and Democracy in the Netherlands* (Berkeley: University of California Press, 1968); and, by the same author, *Power Sharing in South Africa* (Berkeley: Institute of International Studies, 1985).

5. See, especially, Donald L. Horowitz, *A Democratic South Africa?* (Berkeley: University of California Press, 1991); and Arend Lijphart, *Power-Sharing in South Africa*.

6. See, for example, Lynetter Dreyer, *The Modern African Elite of South Africa* (London: Macmillan, 1989), pp. 101–107.

7. Melville Leonard Edelstein, *What Do Young Africans Think?* (Johannesburg: South African Institute of Race Relations, 1972), pp. 95, 107.

8. Ibid., p. 108.

9. Sipho Mila Pityana and Mark Orkin, eds., *Beyond the Factory Floor: A Survey of COSATU Shop-Stewards* (Johannesburg, Ravan Press, 1992), p. 70.

10. Various white right-wing groups, such as the Afrikaner Weerstandsbeweging (AWB), define themselves in ethnic terms, and until recently so did the ruling National Party. But those who are concerned with constitutional engineering to prevent ethnic conflict are usually focused primarily on the politics of the black community.

11. The survey was conducted by MarkData and published by the Human Sciences Research Council. See story in the *Star*, "Mandela, FW Lead Head and Shoulders," April 19, 1992, p. 12.

12. The survey, conducted by the Markinor polling company, utilized a random sample of 1,300 adult black persons in the Pretoria-Witwatersrand-Vereeniging area, Durban, Cape Town, East London, and Port Elizabeth. The results are reported in *Southern African Report*, September 18, 1992, p. 6.

13. See Fatima Meer, *Resistance in the Townships* (Durban: Madiba, 1989), p. 262. A 1985 survey of black political allegiances in metropolitan areas across South Africa, conducted by Mark Orkin, found that support for Buthelezi and Inkatha among Zulu speakers in urban Natal stood at 34 percent and dropped to 11 percent among Zulus living in the Transvaal's urban areas. See Mark Orkin, *The Struggle and the Future: What Black South Africans Really Think* (Johannesburg: Ravan Press, 1986), p. 40.

14. Horowitz, *A Democratic South Africa?* p. 41.

15. Ibid., p. 85.

16. Lijphart, *Power Sharing in South Africa*, pp. 19–20.

17. This presumes, of course, that the transition produces a democratic and workable constitution.

18. For an overview of the work of IDASA, see the organization's journal, *Democracy in Action*.

19. See article in the official ANC journal "The Role of the ANC Branch," *Mayibuye: Journal of the African National Congress*, July 1991, pp. 20–21.

20. See Duma Gqubule, "ANC Economist Spells Out Future," *Star*, February 19, 1992, p. 17.

21. See Lester Thurow, *Head to Head: The Coming Economic Battle Among Japan, Europe, and America* (New York: William Morrow, 1992), pp. 31–33.

22. Ibid.

23. Ironically, the transitional problems of Eastern Europe and the former Soviet Union, where both democracy and economic growth are less likely than in South Africa, may pull so much of the world's available financial assistance that the considerably smaller sum required in South Africa will not be available.

24. See Harold Wilensky, *The 'New Corporatism,' Centralization, and the Welfare State*, Contemporary Political Sociology Series (Beverly Hills: Sage, 1976), p. 23.

25. The National Economic Forum was formed in August 1992 with employer, labor, and government representation.

26. See Wilensky, *The 'New Corporatism,'* pp. 21–23.

Select Bibliography

Books Published Since 1990 on Current South African Politics

Adam, Heribert, and Kogila Moodley. *The Opening of the South African Mind.* Berkeley: University of California Press, 1993.

Barrell, Howard. *MK: The ANC's Armed Struggle.* London: Penguin, 1990.

Baskin, Jeremy. *Striking Back: A History of COSATU.* Johannesburg: Ravan, 1991.

Blumenfeld, Jerome. *Economic Interdependence in Southern Africa: From Conflict to Cooperation.* Cape Town: Oxford University Press, 1992.

Bond, Patrick. *Commanding Heights and Community Control: New Economics for a New South Africa.* Johannesburg: Ravan Press, 1991.

Burman, S., and E. Preston-Whyte, eds. *Questionable Issue: Illegitimacy in South Africa.* Cape Town: Oxford University Press, 1992.

Cawthra, Gavin. *The State of the Police: The South African Police and the Transition to Post-Apartheid Society.* London: Zed Books, 1992.

Chidester, David. *Shots in the Streets: Religion and Violence in South Africa.* Cape Town: Oxford University Press, 1992.

Christie, Pam. *The Right to Learn: The Struggle for Education in South Africa.* Johannesburg: Ravan Press, 1992.

Cock, Jacklyn. *Colonels and Cadres: War and Gender in South Africa.* Cape Town: Oxford University Press, 1991.

Cohin, Robin, Yvonne Muthien, and Abebe Zegeye. *Repression and Resistance: Insider Accounts of Apartheid.* London: Hans Zell, 1990.

Coleman, Keith. *Nationalization: Beyond the Slogans.* Johannesburg: Ravan Press, 1991.

Crewe, Mary. *AIDS in South Africa: The Myth and Reality.* London: Penguin, 1992.

De Klerk, Michael, ed. *A Harvest of Discontent: The Land Question in South Africa.* Cape Town: IDASA, 1991.

Dugard, John et al. *The Last Years of Apartheid: Civil Liberties in South Africa.* New York: Ford Foundation, 1992.

Ellis, Stephen, and Sechaba Tsepo. *Comrades Against Apartheid: The ANC and the South African Communist Party in Exile.* Bloomington: Indiana University Press, 1992.

Everett, David, and Elinor Sisulu, eds. *Black Youth in Crisis: Facing the Future.* Johannesburg: Ravan Press. 1992.

Frederikse, Julie. *The Unbreakable Thread: Non-Racialism in South Africa.* Johannesburg: Ravan Press, 1990.

————. *All Schools for All People: Lessons for South Africa from Zimbabwe's Open Schools.* Cape Town: Oxford University Press, 1992.

Gelb, Stephen, ed. *South Africa's Economic Crisis.* Cape Town: David Philip; London: Zed Books, 1991.

Hansson, D., and D. van Zyl Smit, eds. *Towards Justice? Crime and State Control in South Africa.* Cape Town: Oxford University Press, 1990.

Horowitz, Donald L. *A Democratic South Africa? Constitutional Engineering in a Divided Society.* Berkeley: University of California Press, 1991.

Innes, D., M. Kentridge, and H. Perold. *Power and Profit: Labour, Politics and Business in South Africa—Innes Labour Brief.* Cape Town: Oxford University Press, 1992.

James, Wilmot. *Our Precious Metal: African Labour in South Africa's Gold Industry.* Cape Town: David Philip, 1992.

Lee, R., and L. Schlemmer, eds. *Transition to Democracy: Policy Perspectives.* Cape Town: Oxford University Press, 1991.

Lewis, Stephen R. *The Economics of Apartheid.* New York: Council on Foreign Relations, 1990.

Lipton, Merle, and C. Simkins, *State and Market in Post Apartheid South Africa.* Johannesburg: Witwatersrand University Press, 1993.

Lodge, Tom, Bill Nasson, et al. *All, Here, and Now: Black Politics in South Africa in the 1980s.* New York: Ford Foundation, 1991.

Malan, Rian. *My Traitor's Heart.* London: Bodley Head, 1990.

Mallaby, Sebastian. *After Apartheid: The Future of South Africa.* New York: Random House, 1992.

Maller, Judy. *Conflict and Co-operation: Case Studies in Worker Participation.* Johannesburg: Ravan Press, 1992.

Manzo, Kathryn. *Domination, Resistance and Social Change in South Africa.* New York: Praeger, 1992.

Mare, Gerhard. *Brothers Born of Warrior Blood: Politics and Ethnicity in South Africa.* Johannesburg: Ravan Press, 1992.

Marx, A. *Lessons of Struggle: South African Internal Opposition 1960–1990.* Cape Town: Oxford University Press.

McKendrick, B., and W. Hoffman, eds. *People and Violence in South Africa.* Cape Town: Oxford University Press, 1990.

Minnear, Anthony, ed. *Patterns of Violence: Case Studies of Conflict in Natal.* Pretoria: Human Sciences Research Council, 1992.

Moll, Peter, ed. *Redistribution: How Can It Work in South Africa?* Cape Town: David Philip.

Moss, Glenn, and Ingrid Obery, eds. *South African Review: From 'Red Friday' to CODESA.* Johannesburg: Ravan Press, 1992.

Nattrass, Nicoli. *Profits and Wages: The South African Economic Challenge.* London: Penguin, 1992.

Nattrass, Nicoli, and Elizabeth Ardington, eds. *The Political Economy of South Africa.* Cape Town: Oxford University Press, 1990.

Ohlson, Thomas, and Stephen John Stedman. *The New Is Not Yet Born: Conflict Resolution in Southern Africa.* Washington, D.C.: Brookings Institution, 1994.

Patel, Leila. *Restructuring Social Welfare: The Options for South Africa.* Johannesburg: Ravan Press, 1992.

Pityana, Sipho Mila, and Mark Orkin, eds. *Beyond the Factory Floor: A Survey of COSATU Shop-Stewards.* Johannesburg: Ravan Press, 1992.

Preston-Whyte, E., and C. Rogerson, eds. *The Informal Economy of South Africa: Past, Present and Future.* Cape Town: Oxford University Press, 1992.

Price, Robert M. *The Apartheid State in Crisis: Political Transformation in South Africa: 1975–1990.* New York: Oxford University Press, 1991.

Robertson, M., ed. *Human Rights for South Africans.* Cape Town: Oxford University Press.

Sachs, Albie. *Protecting Human Rights in a New South Africa*. Cape Town: Oxford University Press, 1990.

Schrire, Robert. *Adapt or Die: The End of White Politics in South Africa*. New York: Ford Foundation, 1992.

————, ed. *Wealth and Poverty: Critical Choices for South Africa*. Cape Town: Oxford University Press, 1993.

Seekings, Jeremy. *Heroes or Villains? Youth Politics in the 1980s*. Johannesburg: Ravan Press, 1993.

Slabbert, Frederik Van Zyl. *The Quest for Democracy: South Africa in Transition*. London: Penguin, 1992.

Smith, David M. *The Apartheid City and Beyond: Urbanization and Social Change in South Africa*. London: Routledge, 1992.

Sparks, Allister. *The Mind of South Africa*. London: Heineman, 1990.

Swilling, M., R. Humphries, and K. Shubane, eds. *Contemporary South Africa Debates Series—The Apartheid City in Transition*. Cape Town: Oxford University Press, 1992.

About the Contributors

Amy Biehl was a Fulbright research fellow at the Community Law Centre of the University of the Western Cape; her research focused on the role of women in the democratic transition in South Africa. After graduation from Stanford University in 1988, she worked as a program assistant for the National Democratic Institute in Washington, D.C. Amy Biehl died in August 1993, in Guguletu Township. Her editors wish to express their sense of loss and sorrow.

Colin Bundy is director of the Institute for Historical Research at the University of the Western Cape. He studied at the Universities of Natal and Witwatersrand before proceeding to Oxford as a Rhodes scholar. He has taught in England, the United States, and South Africa. Author of *The Rise and Fall of the South African Peasantry* and (with William Beinort) *Hidden Struggles in Rural South Africa*, he is currently writing a biography of Govan Mbeki.

Jacklyn Cock is associate professor of sociology at the University of the Witwatersrand. Author of *Maids and Madams: A Study in the Politics of Exploitation* and *Colonels and Cadres: War and Gender in South Africa*, she is currently carrying out research on the impact of military-related activities on the ecology of Southern Africa and the problems of psychological trauma among MK cadres. She was a founding member of the Military Research Group in South Africa.

Jeffrey Herbst is assistant professor of politics at the Woodrow Wilson School at Princeton University. He is the author of *State Politics in Zimbabwe* and *The Politics of Reform in Ghana*. In 1992–1993, he was a Fulbright lecturer in politics at the University of Cape Town and the University of the Western Cape, during which time he researched the politics of the economic transition in South Africa.

Herbert M. Howe is the director of African studies at Georgetown University's School of Foreign Service. He has written on Southern African security for *CSIS Africa Notes, Armed Forces Journal, Defence Africa,* and the *Christian Science Monitor* and has contributed chapters on Southern African politics to several books and periodicals. He is currently a coordinator of a USAID-funded diplomatic training program for black South Africans.

Gilbert M. Khadiagala is assistant professor of comparative politics and African studies at Kent State University in Ohio. He has studied in Kenya, Canada, and the United States. He has published articles on African politics and Southern African security and is currently at work on a book manuscript, *Allies in Adversity: The Frontline States in Southern African Security.*

Thomas G. Krattenmaker is professor of law at Georgetown University Law Center, where he teaches constitutional law. For a three-month period during 1991, he was visiting professor of law at the University of Natal (Durban) College of Law in South Africa.

Robert M. Price is a member of the political science faculty of the University of California, Berkeley, specializing in African politics with a primary emphasis on South Africa. He has published widely, including a book on Ghana, a monograph on U.S. foreign policy toward sub-Saharan Africa, and articles on the political economy of change and on South African domestic and foreign affairs. His latest book is *The Apartheid State in Crisis: Political Transformation in South Africa, 1975–1990.*

Stephen John Stedman is assistant professor of African studies and comparative politics at the Johns Hopkins University Nitze School of Advanced International Studies. He is the author of *Peacemaking in Civil War: International Mediation in Zimbabwe, 1974–1980* and (with Thomas Ohlson and Robert Davies) *The New Is Not Yet Born: Conflict and Its Resolution in Southern Africa.* From January to September 1993, he was a senior Fulbright research fellow at the University of the Western Cape, where he studied the politics of negotiation in South Africa.

Peter Vale is director of the Centre for Southern African Studies at the University of the Western Cape. A former president of the Political Science Association of South Africa, he has authored over 200 academic publications. A key contributor to the debate on South Africa's foreign policy in the postapartheid era, he advises the African National Congress on international issues. At present, he is working on a monograph that compares the different foreign policy cultures of the South African state and the African National Congress.

I. William Zartman is the Jacob Blaustein Professor of International Organization and Conflict Resolution and director of African studies at the Nitze School for Advanced International Studies, Johns Hopkins University. He is the author of a number of works on Africa, including *Africa in the 1980's* (with Legum, Langdon, and Mytelka); *Ripe for Resolution: Conflict and Intervention in Africa;* and *Conflict Resolution in Africa* (with Francis Deng). He has also edited and coauthored a number of the works in the SAIS African Studies Library, including *Europe and Africa: The New Phase.*

Index

About the Book

Any multiracial government that emerges in South Africa in the next several years will face a broad and daunting set of challenges. This book considers those challenges, addressing the deep-seated economic, political, and social cleavages that must be overcome if South Africa is to be transformed into a stable democracy.

The authors explore such issues as the need to restructure the economy, integrate both the civil service and the security and police forces, respond to the problems of marginalized youth, achieve political and economic equality for women, and reorient foreign policy and regional relations to provide for equitable growth throughout Southern Africa. Though timely, their analyses are designed to serve also as a lasting guide to an inevitably lengthy period of change.

The SAIS African Studies Library

Botswana: The Political Economy of Democratic Development, edited by Stephen John Stedman

Europe and Africa: The New Phase, edited by I. William Zartman

Ghana: The Political Economy of Recovery, edited by Donald Rothchild

South Africa: The Political Economy of Transformation, edited by Stephen John Stedman

Tunisia: The Political Economy of Reform, edited by I. William Zartman